Space and Society

Series editor

Douglas A. Vakoch, SETI Institute, Mountain View, CA, USA
and California Institute of Integral Studies, San Francisco, CA, USA

More information about this series at http://www.springer.com/series/11929

James S.J. Schwartz · Tony Milligan
Editors

The Ethics of Space Exploration

Springer

Editors
James S.J. Schwartz
Department of Philosophy
Wichita State University
Wichita, KS
USA

Tony Milligan
Department of Theology and Religious Studies
King's College London
London
UK

ISSN 2199-3882 ISSN 2199-3890 (electronic)
Space and Society
ISBN 978-3-319-81975-4 ISBN 978-3-319-39827-3 (eBook)
DOI 10.1007/978-3-319-39827-3

Cover design: Paul Duffield

Printed on acid-free paper

This Springer imprint is published by Springer Nature
The registered company is Springer International Publishing AG Switzerland

For Ian M. Banks, falling always outside the normal moral constraints.

Contents

Chapter 1
Introduction: The Scope and Content of Space Ethics

James S.J. Schwartz and Tony Milligan

1.1 The Purview of Space Ethics

Space ethics epitomizes inter-disciplinarity. Its contributors range from astrobiologists to science fiction authors, from geologists to philosophers, from lawyers to political scientists; and from engineers to planetary scientists. It should come as no surprise, then, that space ethics as a field of inquiry resists a simple, unified description. Rather, it is comprised of a broad spectrum of issues and questions that draw on equally diverse intellectual resources.

On the more "theoretical" side of this spectrum are characteristically normative- and meta-ethical questions, i.e., questions related to the construction, standing and evaluation of ethical *theories*: Does the space environment (including the solar system and beyond) contain anything of inherent value (i.e., anything that is valuable for its own sake)?[1] Or is space a mere instrument available for the satisfaction our preferences? What is the moral status of our relationships to various aspects of the space environment—e.g., do we have an ethical obligation to respect

[1]The editors have favored 'inherent' value, in line with a familiar distinction in analytic ethics between 'inherent value' (possessed by that which is of value in its own right) and 'intrinsic value' (possessed by that which is of value to a sentient being without consideration of any further advantage). However, given that these terms are often used synonymously, particularly when ethicists and scientists collide, this favoring of 'inherent' has not been enforced by editorial fiat throughout the volume. Contributors have been left to deploy their preferred terminology.

J.S.J. Schwartz (✉)
Department of Philosophy, Wichita State University, 1845 N Fairmount, Wichita, KS 67260, USA
e-mail: james.schwartz@wichita.edu

T. Milligan
Department of Theology and Religious Studies, King's College London, London, UK
e-mail: tony-milligan@outlook.com

J.S.J. Schwartz and T. Milligan (eds.), *The Ethics of Space Exploration*, Space and Society, DOI 10.1007/978-3-319-39827-3_1

or constrain our activities on entities such as asteroids, comets, moons, or planets? If extraterrestrial life (including microbial life) is discovered, would it fall under the scope of moral consideration? And if so, in what way? And for what reasons? And to what degree? In short, how should the consideration of the space environment impact upon the way in which we *reason* about *what to do* and about *what matters*?

Meanwhile, on the more "practical" side of the spectrum, there are a variety of questions about the ethical evaluation of existing and proposed activities in space:

1. Can national and global expenditures on space exploration be justified? Is existing support adequate, insufficient, or superfluous? How should this support be divided between human and robotic exploration?
2. What are the risks associated with various forms of space travel, including long-duration spaceflight? Are space travel participants given adequate information about these risks, and is their assessment of them sufficiently objective?
3. To what extent should we preserve pristine space environments, such as asteroids or planetary surfaces? Are sites potentially home to extraterrestrial life (or traces of past life) more worthy of preservation? How diligently must we work to avoid contaminating extraterrestrial sites with terrestrial microbes, some of which might survive long periods of exposure to vacuum? Why is any preservation warranted in the first place—to protect opportunities for scientific research? Or because, e.g., asteroids or planetary surfaces, or extraterrestrial life forms, are inherently valuable and hence worth preserving for their own sake?
4. What is the most fair and effective way to regulate particularly "popular" locations in space, e.g., low-Earth orbit (LEO) and geostationary orbit (GEO)? Orbital slots in LEO are particularly useful for Earth observation satellites, and LEO is the dominant milieu of human spaceflight. Meanwhile, GEO is particularly useful for global telecommunication satellites. Should access to positions in Earth orbit be permitted on a first-come, first-served basis, or should access to Earth orbit be subject to some kind of social justice constraint?
5. Debris from nearly 60 years of activity in space poses an increasing hazard to both human and remote operations in Earth orbit. What responsibilities do we have to limit the production of this debris? Are we obliged to "clean up" this debris if we can develop the requisite capabilities?
6. Should property rights be granted to those interested in developing space resources, e.g., to corporations such as Planetary Resources, which are interested in extracting mineral resources from asteroids? Should the granting of these rights be made on a basis of first-come, first-served, or should there be an equitable sharing of the resources from space? Would it ever be permissible to terraform a planet, i.e., to use geophysical engineering to turn a previously uninhabitable planet into one that is suitable for human settlement?
7. What kinds of challenges will denizens of space colonies and settlements face? What form of governance or social organization would maximize colonists' security and their personal liberty in an *extremely hostile* environment where basic resources, such as water and air, must be *manufactured?*

As the reader will see, most of these questions—"theoretical" and "practical" alike—are treated in the contributions to this volume. And for a variety of answers to them we point you to the subsequent chapters.

What we wish to emphasize at present is that none of the above questions fall under the exclusive remit of any one scientific, or philosophical (or other liberal arts) discipline. For instance, the "theoretical" task of constructing an ethical theory for conduct in space should draw not only on discussions in philosophy (viz., normative ethics), it should also draw on the space sciences—astrobiology, astronomy, planetary science, etc. After all, an adequate ethic for space should be designed with some minimal account of what the space environment is comprised of. Similarly for the specific question of the moral status of extraterrestrial life, which implicates not only astrobiology and normative ethics, but also chemistry and biology in the sciences, and in philosophy, bioethics, environmental ethics, philosophy of science, and the philosophy of language.

Regarding the more characteristically "practical" questions, resolving those identified in (1) relies critically on the space and social sciences. The space sciences for limning the possibilities of exploration and the expansion of scientific knowledge; the social sciences (anthropology; economics; philosophy; sociology) for assessing and predicting how the course of space exploration will impact upon society. After all, the value of scientific research more generally, and space research in particular, is not often self-evident—its manifestation requires a *human* perspective. Detailing the risks implicated in (2) requires extensive biological and medical knowledge about the effects of reduced- and micro-gravity environments on living organisms. Thus any ethical assessment of the risks posed to space travel participants would not only be an exercise in bioethics and business ethics, but also an exercise in the life sciences. In addition to many of the disciplines already mentioned, the questions identified in (3)–(6) pertain to the law and regulation of space, and thus call on counsel from space lawyers and policymakers. And finally, the task of how to maximize liberty and security in space settlements (7) raises questions of interest to engineers, political philosophers, political scientists, psychologists, and sociologists.

Neither our list of 'big questions' nor our list of relevant disciplines is meant to be exhaustive, but we hope that the reader is left with the impression that space ethics presents a unique and engaging setting for the fruitful interchange of ideas from a diverse array of scientific and liberal arts perspectives. To this end, the contributions to the volume are accessible and state-of-the-art overviews of several of the major issues in space ethics, and should provide readers with an authoritative introduction to the key issues in the field. However, since a history of space ethics is not the particular subject matter of any contribution, we find it helpful here to provide a brief overview of the development of space ethics as it pertains (broadly) to the contributions to this volume.

1.2 Formation of the Discourse

Space ethics, understood in the above sense (and rather like the space program itself) experienced something of a false start with a handful of early publications in the late 1970s and 1980s, primarily as an offshoot of the growth in environmental ethics and with key contributors drawn from the latter (notably, Holmes Rolston and J. Baird Callicott, but work by Wendell Berry and James Lovelock also had some clear ethical dimensions). The most important of these early contributions appeared in Eugene Hargrove (ed.) *Beyond Spaceship Earth*: *Environmental Ethics and the Solar System* (1986) which, although tentative, exploratory and of uneven quality, still remains an important reference point on questions of inherent value and the permissibility of terraforming. An ongoing feed-in from environmental ethics remains clear in several contributions to the present volume, and the debates covered in the Hargrove collection have continued intermittently, through 'outlier' articles about terraforming in philosophical ethics journals and collections, although these matters remain situated at some distance from the policy agenda and from the main debates in philosophical ethics (for example, McKay and Davis 1989 and McKay 1990).

While in many respects a disadvantage, the distance from policy during the early formative stages of the discourse has sometimes had its compensations. Most notably, the work on space ethics that finally began to emerge as part of a more cohesive discourse from around 1999 onwards, was not excessively concerned with the immediate priorities of the Shuttle Program which might have produced an interestingly different sort of space ethics geared towards other priorities such as engineering-related risk, resource prioritization and the character required of pioneering space agents (Pinkus et al. 1997). Instead, these considerations appear in this volume as part of a broader cluster of issues. Nor has the discourse succumbed to the remoteness from lived ethical experience, which, from time to time, has afflicted contemporary philosophical ethics making it difficult for the latter to establish pathways towards research impact. Instead, the contemporary discipline of space ethics has been shaped by two dominant, lines of discussion both of which have at least some institutional connections to NASA and the European Space Agency: one which was opened up in the US at the end of the 1990s following on from the rapid growth of astrobiology and exploring the ethical issues raised by the latter and its connection to broader societal issues, and another which was opened up in Europe, at much the same time, on areas where ethics, policy and space law intersect (the most obvious of these being matters of commercialization, the distribution of opportunities and responsibilities in nearby regions of space) as well as the rationale for space exploration.

The astrobiology and society discussion, as a serious output-oriented affair, began at a workshop on the Societal Aspects of Astrobiology held at the Ames Research Centre in 1999 and a AAAS workshop in Washington in 2003, culminating a decade later in the first major contemporary volume dealing with space ethics, Constance Bertka (ed.) *Exploring the Origin, Extent and Future of Life*:

Philosophical, Ethical and Theological Perspectives (2009), together with a proposed "Roadmap of Societal Issues," drafted at a workshop hosted at the SETI Institute (Rummel et al. 2012). The roadmap mimicked (and also picked up on the endorsement of social and ethical issues) in NASA's 'Astrobiology Roadmap' (Des Marias et al. 2008) but never quite secured the institutional support given by NASA to the latter, for understandable reasons. Even so, it set a pattern for the inclusion of ethical matters with the search for life and debate about its possible value. The five roadmap goals were formulated in terms which were, from the outset, ethical-friendly:

> (A) Explore the range and "complexity" of societal issues related to how life begins and evolves.
> (B) Understand how astrobiology research relates to questions about the significance and meaning of life.
> (C) Explore the relationships of humans with life and environments on Earth.
> (D) Explore the potential relationships of humans with "other" worlds and types of life.
> (E) Consider life's collective future—for humans and other life, on Earth and beyond.
> (Rummel et al. 2012, p.959)

One notable feature of this US side of the discussion, from its beginnings as an offshoot of astrobiology, has been the interest in societal impact consistently shown on matters of religious commitment. This has carried over into the most recent edited volume on the area edited by Chris Impey, Anna H. Spitz, and William Stoeger, *Encountering Life in the Universe*: *Ethical Foundations and Social Implications of Astrobiology* (2013), and into NASA-funded research work at the Princeton Centre for Theological Inquiry.

The European discussion has emerged partly out of the drafting of the Pompidou Report *The Ethics of Space Policy* (2000) presenting the outcome of meetings and discussions by a joint UNESCO/European Space Agency working group on the "Ethics of Outer Space" held over the previous 2 years. Its tone and content were very different from the US discussion, with an opening declaration that 'Earth and Space are not ours. They are treasures, real and symbolic, which we owe to ourselves to safeguard for our descendants.' (Pompidou 2000, ii). Jacques Arnould, one of the key contributors to the report (and to the volume here), then produced the first systematic exposition of issues in the field the following year, in *La Seconde Chance d'Icare* (2001), which appeared, translated and expanded, as *Icarus' Second Chance*: *The Basis and Perspectives of Space Ethics* (2011), arguing that the most appropriate form of frontier ethic for space was a form of humanism. (In the sense of an ethic that places human identity at its heart, rather than in the sense of something that requires anthropocentric judgements about the value of the non-human.)

With the emergence of these two interweaving discussions, the discipline can now reasonably claim to have arrived at a stage of early consolidation, a point where key lines of argument and dispute have become clear: some contributors argue that everything 'out there' is open to our appropriation, others urge caution out of a sense that humans matter but we are not all that matters. Others urge caution for the sake of humans and because access to space and all it holds is unevenly (and perhaps unfairly) distributed. Whatever position is taken on these questions, since the publication of the Roadmap and the Pompidou Report, the field has attracted a growing body of academics based in the disciplines of philosophy, the philosophies of biology and science, geoethics (where geology and ethics collide), astrobiology, anthropology and space law. What has marked the field during this phase of consolidation is a combination of two things: a high level of interdisciplinary integration spanning the humanities and relevant disciplines of science, and some tentative degree of institutional connectedness in terms of links with NASA, the European Space Agency and bodies such as COSPAR (the Committee on Space Research established back in the 1950s after the launch of Sputnik and a major player in the discussions on policy and space law).

Where we find ourselves today is also at something of a crossroads with NASA issuing a new *NASA Astrobiology Strategy* (2015) document to replace its earlier "Astrobiology Roadmap," removing much of the societal content of the latter, with the role of ethics still acknowledged but pushed into the margins, consigned to a single paragraph in a document almost exclusively given over to science conceived of in terms which are functional to the discovery of life (Hayes 2015, p. 159). This too is understandable for all sorts of reasons. There is a reasonable concern that ethics may be a source of unhelpful constraints which could stand in the way of the emerging space economy and (on a less commercial note) there is a widespread view that we may now have a good idea of where to look for and discover life elsewhere in the Solar System if there is any to be found. Astrobiology is, to some extent, gearing up for a possible first discovery and all sorts of add-on issues have tended to be pushed into the long grass in order to concentrate upon what is (appropriately or otherwise) deemed to be the central task. Paradoxically, space ethics has been pushed out to the margins at a time when (with the prospect of possible discovery and the emergence of a space economy) it has matured to the point of being able to contribute significantly to the discussion on human activity in space and at a time when its contribution on the ways in which we may legitimately explore may be particularly salient.

There are also various institutional counter-trends which put ethics much closer to the heart of this discussion. While the *NASA Astrobiology Strategy* (2015) turned away from ethics, there has been a marked strengthening of the interest in ethics on the European side with an initiative to establish a European Institute of Astrobiology broadly along the lines of the earlier 'Roadmap of Societal Issues' amid a growing sense of positioning or preparing for discovery, if there is microbial life out there. And NASA too has become involved in the funding of research on the societal impact of astrobiology through Princeton's Centre for Theological Inquiry, again taking the route of ethico-theological reflection on what it would mean to discover life (of any sort) elsewhere in the universe. It seems that these are matters which simply cannot be sent permanently into exile. And the quality of the ethical deliberation itself has improved dramatically from its earliest tentative beginnings, with the Impey, Spitz and Stoeger volume appearing in 2013; a special edition of the journal *Space Policy* on ethics appearing in 2014; a volume edited by Jai Galliot (2015); and a series of three volumes issued by Springer and edited by

Charles Cockell (2015a, b, 2016) on the politics of space settlement and the ethico-political theme of liberty. On the side of monographs, a short book by one of the editors, Tony Milligan, *Nobody Owns the Moon: The Ethics of Space Exploitation* was released in 2015 and further monographs by Gonzalo Munévar and Mark Lupisella (also contributors to the present volume) are anticipated shortly. The discourse has already reached the 'usual suspects' stage at which any volume or special edition of a journal is expected to contain articles by some subset of established key contributors, and the present volume certainly has those. But it has also been our aim to reflect the fact that space ethics is a growing discipline with new voices, and multiple dialogues going on at one and the same time. A balance has therefore been struck, or at least aimed at.

1.3 An Overview of the Volume

The papers gathered here try to shed light on many of the key established areas as well as exploring some new lines of analysis. The opening section sets the scene by placing space ethics in the context of human history, deep imperatives and culture. The lead paper, from the well-known science fiction author Stephen Baxter, looks at responses in science fiction to Gerard O'Neil's influential text (1976), an ecologically-driven proposal to deal with energy and population problems through the construction of mass habitats in free space. Baxter's critical assessment of these surprisingly detailed and cogent literary responses to O'Neil considers not only the multiple impracticalities of the initial proposal but also the fundamental imperatives behind mankind's growth and the desire to reach outwards. In Chap. 3 Gonzalo Munévar flips over and, instead, interrogates O'Neill's environmentalist critics in an attempt to renew the vision of artificial worlds offered by the latter, without the initial design difficulties and with underpinnings from evolutionary biology and an appeal to survival concerns in the face of extinction threats.

In Chap. 4 on 'Agonal Conflict and Space Exploration' Eleni Panagiotarakou shifts the theme toward the competitive dimension of space exploration and the ways in which the Cold War tensions between the USSR and the USA, from which the space program can be seen to parallel the ancient *agon*, a rivalry in which the enemy is to be surpassed in excellence but not destroyed. For Panagiotarakou, in the aftermath of the USA/USSR conflict, what has emerged on the USA side is not harmony but a problematic tension of a different sort, one between NASA and the private sector. Struggle of a different sort figures prominently in Chap. 5 where Chris Yorke explores Frederick Turner's 'Frontier Thesis' that the character of a people is determined by the severity of the obstacles which they face. The claim is that without the rigors of a frontier experience human culture is liable not to consolidate but to stagnate. Yorke upholds the 'frontierist' position against the 'consolidationist' position and argues that human culture on Earth will ultimately depend upon our exploring the utopian potential of space.

Shifting from the historical and cultural context directly to the link-up between space and hard core ethical theory, Mark Lupisella's contribution in Chap. 6 considers the case for a 'cosmological ethic' which stresses connectedness and relationality as a grounding for talk about inherent value. This draws upon a proposal for a special cosmological or 'cosmocentric' resetting of ethics that has been around continuously from the opening discussions in the 1980s but whose content has always been difficult to pin down. Lupisella gives an overview of how his forthcoming book on space ethics attempts to tackle the problem. In Chap. 7 Schwartz picks up on the value theme and challenges the idea that we know enough about the space environment to say for sure just what is and is not of value. An attempt to fix this prematurely may cut us of from novel reasons for planetary protection that might otherwise appear. Shifting the weight of assessment towards character, an option found instead in virtue ethical approaches (and aligned approaches, such as those of Yorke and Milligan), also does not avoid the problems that arise from this epistemic shortfall and raises the classic virtue ethical problem of being insufficiently guiding in our actions and choices. By contrast, Seth Baum's contribution in Chap. 8 adopts a firmly consequentialist approach to matters and sets out to explain why measures to deal with global risk may be more important than maximizing goods elsewhere. Baum also explores the possibility that a strict consequentialism might lead us to prioritize the interests of a more advanced species rather than those of humans (and the possible implications that this line of thought may have for our current treatment of terrestrial non-humans). In Chap. 9, Milligan reaffirms Schwartz's more pluralist approach with a similar precautionary resistance to any attempted foreclosing of deliberation about the kind of ethic that might be appropriate to sustained life in space. He attempts to strengthen this caution through a rejection of any idea that we should be trying to establish an unchanging set of foundations for space ethics (irrespective of whether these foundations are thought of in consequentialist terms or in some rival terms). The claim is that while some rudimentary side-constraints concerning certain sorts of harm may always be in place, they are too slight to perform this task. In short: the right kind of ethic for terrestrial life and even for early settlement might not be the right ethic for everywhere and at all times although every ethic will contain at least some important (non-foundational) features in common. What kind of ethic might then fit remote contexts is something that we are poorly placed to judge. (As with Schwartz, epistemic disadvantage plays a role.)

Chapters 10 and 11 shift us into two contrasting discussions about the nature of our humanity: the formation of our *way of being;* and how space exploration and expansion may impact upon identity. For Francesca Ferrando, in 'Why Space Exploration must be Posthuman,' space calls upon us to simultaneously move beyond consequentialist technologically-focused perspectives and beyond conceptions of the human. For Jacques Arnould, in 'An Urgent Need to Explore,' space opens up new and undogmatic ways of being and conceiving of the human with exploration itself taken to be a necessary part of human existence. (Here, Arnould picks up on a familiar claim about an imperative to explore, introduced in Baxter's and Munévar's chapters and appearing elsewhere in the volume.)

Chapters 12–15 deal with the contested areas of value theory and the non-human, looking at the issues of planetary protection and microbial value and opening with a paper by Charles Cockell, the leading exponent of inherent or non-instrumental value of extraterrestrial microbes. Cockell argues that although the protection of individual microscopic organisms is impractical in the course of planetary exploration, we can nonetheless develop an ethic of respect for communities of microscopic life, valuing them for more than the uses to which we put them, and avoiding the wanton destruction of other life and biospheres in the pursuit of our own objectives. In Chap. 13 Anna Frammartino Wilks returns to the theme of value and 'cosmocentric ethics' of the sort explored by Lupisella and attempts to give foundations for the latter by appeal to Kant's distinction between *self-organizing* beings (life forms) and *self-legislating* beings (moral agents) as a non-arbitrary basis for claims about value. More specifically, Wilks argues for a '*weak* cosmocentric ethic,' *cosmocentric* in the sense that it acknowledges the priority of interest claims for all valuable beings, *weak* in the sense that it does not attribute inherent value to the universe itself, over and above the beings within it.

Chapter 14 returns to the issue of inherent value with a critique of the latter by Kelly C. Smith. In 'Why Microbes lack Inherent Value,' Smith develops and expands a position that he first advanced in the Constance Bertka volume by challenging what he calls 'Mariomania,' the view that our engagement with Mars ought to be dictated by the interests of microbial life forms if any of the latter are found. This view is exemplified by Carl Sagan's call to accept that Mars would belong to the Martians, if there were any, and even if they were microbial. Rather than focusing upon the familiar arguments which are found in Cockell, in the environmentalism-influenced literature and in Smith's earlier papers, he instead challenges the ethical underpinnings of such an approach through a critique of its appeal to inviolable principles concerning life-forms, and by appeal instead to a form of ethical pragmatism that highlights the considerable opportunity costs of foregoing the human exploitation of Mars. In this way Smith touches upon the deep background intuition that sometimes motivates scientific opposition to inherent value arguments. In Chap. 15 Sean McMahon shifts the ground for planetary protection away from matters of microbial value towards the appeals for and against the terraforming of Mars which draw upon aesthetic considerations. McMahon argues that Mars may offer distinctively Martian forms of beauty that we are not yet in a good position to fully appreciate but which might be brought more fully into the public consciousness by future exploration. In a cautiously formulated claim, McMahon suggests that recognition of such beauty would provide at least a defeasible reason for caution about terraforming. As elsewhere in the protection literature, reasons for caution do not necessarily translate into any comprehensive 'hands-off' attitude.

The final cluster of papers tackle matters of legality and risk. In Chap. 16 'The Way to Eden,' Christopher Newman points to the consensus about damage caused by existing space activities and looks towards the possibility of a more sustainable approach. Newman argues that any new regulatory framework should embed a commitment to environmental protection and that this should be at the forefront of policy discussions. The thought here does not simply concern the legal and ethical issues raised by the prospect of a human-crewed mission to Mars, but rather the template for all long duration space flight. Behavioral norms and regulation are taken to go together in the paper rather than the former constituting the need to circumvent the latter. From environmental protection, Paul Graves moves into the territory of human risk in Chap. 17 considering the danger of high velocity fallback accidents in the case of nuclear powered probes. Such accidents could spread radioactive material across a broad area and although fallback accidents are uncommon, Graves' utility analysis suggests the likelihood of an unacceptably high expected loss of life, one which would not be adequately justified by appeal to the expected returns from science probes. Consequentialist and deontological critiques of current practices are therefore accepted as broadly correct with the upshot that the associated deep space programs should not be abandoned but modified, to properly acknowledge the dangers.

The final chapter, by Frans von der Dunk, 'Shaking the foundations of the law: some legal issues posed by a detection of extra-terrestrial life,' takes us to the widest reaches of our aspirations in space: the discovery of other intelligent life forms and the legal issues that this would pose. If such beings possessed intelligence of a lesser sort and had no concept of law, would it be legally permissible to treat them as objects of our law, with rights requiring advocacy (rather like animals) instead of agents who might speak for themselves? If they had a similar intelligence and their own concept of law, then whose law should be upheld? The argument is made by von der Dunk for a compromise 'meta-law', arranging the respective spheres of application of human-made law and the comparable extra-terrestrial legal system. However, if they possessed greater intelligence (a scenario considered already in the Baum paper) we might instead become the 'object' of their system of 'law.' These considerations are used by von der Dunk to problematize our sense of the stability of law in the face of a possibility of discovery that space exploration poses.

The aim throughout the book has been to provide a series of ethical encounters that will help to set out, illuminate and stimulate key arguments and help to mark the arrival of space ethics at a new stage in its development. The editors are indebted to the initial encouragement of the Space and Society series editor Douglas A. Vakoch for helping to shift the idea for a volume from email exchanges to something more concrete; to Alessia Valdarno and Ramon Khanna at Springer for ongoing guidance and support; to the contributors for their patience and diligence, to Mukhesh Bhatt for providing valuable commentary on a cluster of these papers at an Ethics of Space Exploration workshop held at the University of Hertfordshire in 2015 and to numerous friends and colleagues across the space community for their interest and commitment to the importance of ethical deliberation.

References

Arnould, J. (2011). *Icarus' second chance: The basis and perspectives of space ethics*. Vienna and New York: Springer.

Bertka, C. (Ed.). (2009). *Exploring the origin, extent and future of life: Philosophical, ethical and theological perspectives*. Cambridge: Cambridge University Press.

Cockell, C. (Ed.). (2015a). *The meaning of liberty beyond earth*. Heidelberg, New York and London: Springer.

Cockell, C. (Ed.). (2015b). *Human governance beyond earth*. Heidelberg, New York and London: Springer.

Cockell, C. (Ed.). (2016). *Dissent*. Heidelberg, New York and London: Revolution and Liberty Beyond Earth. Springer.

Des Marias, E. J., et al. (2008). The NASA astrobiology roadmap. *Astrobiology, 8*, 715–730.

Galliott, J. (2015). *Commercial space exploration: Ethics, policy and governance*. Farnham: Ashgate.

Hargrove, E. (Ed.). (1986). *Beyond spaceship earth: Environmental ethics and the solar system*. San Francisco: Sierra Club Books.

Hays, L. (Ed.). (2015). NASA Astrobiology strategy. Retrieved from https://astrobiology.nasa.gov/uploads/filer_public/01/28/01283266-e401-4dcb-8e05-3918b21edb79/nasa_astrobiology_strategy_2015_151008.pdf.

Impey, C., Spitz, A. H., & Stoeger, W. (Eds.). (2013). *Encountering life in the universe: Ethical foundations and social implications of astrobiology*. Tucson: University of Arizona Press.

McKay, C. P. (1990). Does mars have rights? An approach to the environmental ethics of planetary engineering. In D. MacNiven (Ed.), *Moral expertise* (pp. 184–197). New York: Routledge.

McKay, C. P., & Davis, W. (1989). Planetary protection issues in advance of human exploration of Mars. *Advances in Space Research, 9*, 197–202.

O'Neil, G. K. (1976). *The high frontier: Human colonies in space*. Burlington: Apogee.

Pinkus, R., Hummon, N. P., Shuman, L. J., & Wolfe, H. (1997). *Engineering ethics: Balancing cost, schedule, and risk-lessons learned from the space shuttle*. Cambridge: Cambridge University Press.

Pompidou, A. (Ed.). (2000). The Ethics of Space Policy, UNESCO.

Rummel, J. D., Race, M., & Horneck, G. (Eds.). (2012). Ethical Considerations for Planetary Protection in Space Exploration. Astrobiology, 12, 1017–1023.

Part I
The Cultural and Historical Context of Space Ethics

Chapter 2
Dreams and Nightmares of the High Frontier: The Response of Science Fiction to Gerard K. O'Neill's *The High Frontier*

Stephen Baxter

2.1 Introduction

> 'Centurion, the cylinder is nearly three thousand miles long.'
>
> 'Three *thousand*—'
>
> 'That is more than the diameter of Luna, sir. The end hubs alone could swallow a small moon. The land area within must be similar to that of the whole of Asia … The question is, of course, who would live in such a structure—'
>
> 'I can tell you that, *optio*, 'Quintus said. '*That's* where the emperor will be. And the very rich. Living off the huge rivers of goods that flow between the worlds.'
>
> 'An emperor become a god,' Titus said. 'I wonder how you could ever get rid of him?'
> Quintus grinned back. 'Good question, Titus.'
> (Baxter 2014, 268–269)

This paper concerns the imaginative response of writers of science fiction (SF) to the proposals for space colonisation developed by Gerard K. O'Neill and co-workers in the 1970s (O'Neill 1976a). (Elsewhere in the present volume Munevar explores criticisms of the O'Neill scheme from a wider audience.)

O'Neill is associated with large space-habitat designs such as the 'O'Neill cylinder' (ibid., 64ff). O'Neill's work is however largely sociological in intent rather than technological. He uses space-habitat designs as stepping-stones in a vision of a progressive future for mankind in the longer term, with small communities 'homesteading' the asteroids in an analogy with the American frontier experience. This is a very science fictional scenario, but O'Neill claimed that his scheme was based on economic and engineering logic, not on SF readings. And as will be seen, the reception of O'Neill's ideas by the science fiction field has been a

S. Baxter (✉)
British Interplanetary Society, 27-29 South Lambeth Road,
London SW8 1SZ, UK
e-mail: sbaxter100@aol.com

J.S.J. Schwartz and T. Milligan (eds.), *The Ethics of Space Exploration*,
Space and Society, DOI 10.1007/978-3-319-39827-3_2

complex one, and not always positive. The best of these fictional responses serve as thought experiments on the plausibility of the O'Neill space colonisation scheme, and the possible reality of human life within its parameters.

2.2 The Sociology of O'Neill and Precursors in Science Fiction

2.2.1 The O'Neill Prospectus

O'Neill's studies are associated with designs of large space habitat. Indeed such designs have become part of the imaginative furniture of the future; for example an O'Neill cylinder was featured, without explanation, in a brief scene towards the end of the movie *Interstellar* (2014, dir. C. Nolan). Such colonies regularly feature in prose fiction too, such as in Iain M. Banks' *The Algebraist* (2004). In the year 4034 AD, in a system called Ulubis twenty thousand light years from Earth, Hab 4409 is an O'Neill cylinder fifty kilometres long, 'a giant, verdant city rolled up into a spinning tube' (Chapter Three).

However, in O'Neill's scheme, large near-Earth habitats would be merely the first stepping stones into space. Their initial economic justification would be to sustain large populations of workers who would build orbital solar power stations (SPS), the output of which would be sold back to the Earth. Arguing from a premise that 10,000 workers in space would be needed to kick-start a significant industrial presence there (O'Neill 1976a, 116) O'Neill proposed as a model starter colony his 'Island One', a sphere ~500 m in diameter, rotating twice per minute to provide Earth-equivalent gravity. This would be constructed largely from lunar resources and would host 10,000 people living at urban population densities. Island Two would be an expanded version with a population scaled up to 140,000 people, and O'Neill's Island Three (ibid., 64ff) was to be a pair of rotating cylinders each 32 km long and hosting a population of 20 million (ibid., 69).

Once humanity was established outside the gravity well, a wider strategy would unfold, with the islands used as bases for further expansion into space. For the space colonies to achieve economic independence from Earth, they would need an extraterrestrial supply of compounds of carbon, nitrogen and hydrogen—materials not available from the moon, but from the asteroids (ibid., 251). Thus O'Neill imagined small groups of people equipped with relatively simple spacegoing technology able to set off from the first islands to 'homestead' the asteroids (ibid., 233), making a living by selling essential materials back to the space colonies.

The consequent transformation in the fortunes of humanity would be dramatic. O'Neill predicts a rapidly bootstrapping human expansion into the solar system, with an extraterrestrial population measured in billions within a few decades (ibid., 260), and a growth of economy and exploitation that would see the resources of the solar system consumed in a few thousand years (ibid., 247).

This was a very science-fictional plan. But O'Neill's visions do not, however, seem to have been influenced by prior science fiction.

2.2.2 Space Colonisation in SF Before O'Neill

O'Neill makes clear that the source of his inspiration was social, not technological: 'Often people have asked why I picked as our first question: "Is a planetary surface the right place for an expanding technological civilisation?" There is no clear answer, save except to say that my own interest in space as a field for human activity went back to my own childhood, and I have always felt strongly a personal desire to be free of boundaries and regimentation' (ibid., 279). While he claims to have read SF as a child (ibid., 60) he recalled no mention of space habitats as an arena for human civilisation, as opposed to moons and planets: 'As a reader of science fiction in childhood, I gained no clue that the future of mankind lay in open space rather than on a planetary surface. Later … logic and calculation forced me to that conclusion' (ibid., 60). He was directed to Tsiolkovsky's fiction, for example, only after his own first designs had been published. He would write, 'In a round-table TV interview, Isaac Asimov and I were asked why science-fiction writers have, almost without exception, failed to point us towards [space colonies]. Dr. Asimov's reply was a phrase he has now become fond of using: "Planetary chauvinism"' (ibid., 35).

However there were indeed precursor works depicting space stations and colonies dating back more than a century, many of which foreshadowed elements of O'Neill's studies. A comprehensive though somewhat dated survey of this SF subgenre was given by Westfahl (2009). These works were not developed in isolation; SF has always attracted a strong community, with readers and writers following each others' work and elaborating on and critiquing shared ideas. In addition there has been a constructive dialogue with philosophers, engineers and others working in the field.

It was in fact in an SF novel, by Konstantin Tsiolkovsky (1857–1935), a Russian scientist and writer, that the fundamental principles of space colonisation were first set out in a coherent fashion: that is, the use of abundant solar energy and other extraterrestrial resources to sustain a large, expansive human future beyond the Earth, the basic scheme that would underpin O'Neill's prospectus. *Vne Zemli* (*Beyond the Planet Earth*) (1920), set in the year 2017, features liquid-fuelled rockets that reach the moon in 4 days (chapter 3), the collection of solar energy in space (chapter 36), spin gravity (chapter 15), and large colonies in cylindrical sunlit 'greenhouses' positioned in geosynchronous orbit (chapter 29). The moon is rather dismissed as a source of raw materials for new colonies—but a near-Earth asteroid, as it would now be called, is prospected (chapter 51). In all this was a remarkably prescient and coherent vision of a human expansion into space.

As to the specific design of space habitats, it was in the famous *Collier's* magazine articles of the 1950s by von Braun and others (Ryan 1952) that the first coherent post-World War II plan for space travel with soundly based engineering was publicised, as developed by the engineers who would go on to drive the US space programme in the 1960s and beyond. And the centrepiece of the study is a wheel-shaped Space Station. It cannot be denied that Von Braun's wheel design has become imprinted on the popular imagination, as 'the' classic space station architecture. The movie *2001: A Space Odyssey* (1968, dir. S. Kubrick) shows perhaps the most famous fictional wheel-in-space, Space Station V, at which Dr Heywood Floyd transfers from an Earth-to-orbit shuttle to a lunar ferry.

But many decades earlier, some SF writers had been led through the engineering logic of spin gravity to anticipate the 'O'Neill cylinder' (Island Three). Williamson's 'The Prince of Space' (1931) is a pulp-fiction saga of the attempted invasion of Earth by plant-like vampire Martians. The eponymous rogue's habitat is a spinning cylinder 5000' (1520 m) in length and diameter, and home to 5000 people. It is an authentically realised O'Neill habitat: 'It gave Bill a curious dizzy feeling to look up and see busy streets, inverted, a mile above his head. The road before them curved smoothly up on either hand, bordered with beautiful trees, until its ends met again above his head' (Chapter 3).

Just as decades of precursor SF prepared humanity for O'Neill's visions, so responses to his schemes would be expressed in fictional form after his first publication.

2.3 Utopias on the Space Frontier

2.3.1 First Reactions

It is easy to see why O'Neill's ideas struck a chord with space advocates. O'Neill's work produced the first detailed post-Apollo space colony designs to be based on plausible modern materials and technologies. He devised a fresh synthesis by integrating old ideas, such as the lunar mass driver, with new results such as the post-Apollo analysis of lunar rocks and their mineral content and potential for use as construction materials. The idea of selling solar energy to the Earth was a new justification for large stations in orbit. His results were analytical, numerate, and compellingly argued.

Not only that, O'Neill published at a time when space exploration had only recently revealed the worlds of the solar system, notably the moon and Mars, to be much less promising in terms of colonising potential than had once been thought: 'When Mariner IV looked on the face of Mars and found only a dead world … a frontier died that afternoon,' space advocate and SF writer Jerry Pournelle would write (1979, 1). Now a vision of habitable destinations in space itself, as opposed to on those disappointing worlds—recall that O'Neill used the term 'islands' to describe his first colonies—would evoke a response from space dreamers of all kinds.

An immediate and generally enthusiastic first response to the O'Neill prospectus was a two-part anthology edited by Pournelle (1979–1982) consisting of original stories and reprints dating from 1975 to 1979. These roughly track through the steps of O'Neill's proposed advance into space. 'Spirals' by Niven and Pournelle is about a race to complete the building of the first O'Neill colony, called the Construction Shack: 'I was a tiny chick in a vast eggshell' (ibid., 36). As the economy on Earth collapses, the US administration steadily cuts back on the station's funding, until the crew convert the station into a ship and sails out to the riches of the asteroid belt. The conflict between the visionary spacers and the short-sighted Earthbound and their governments, called 'downers' here, is characteristic of these stories—and in such polemic pieces the 'downers' are portrayed entirely negatively. Pournelle's own 'Bind Your Sons to Exile' is about the first fully fledged asteroid mine, but just as in 'Spirals' opposition from sceptics on the ground starves the project of funding: '"Boondoggle" was the kindest word they had for us' (ibid., 256).

As for life in the habitats themselves, perhaps the most interesting of the stories here is Sheffield's 'Transition Team', in which a 3000-person O'Neill colony is having significant trouble with its young people. The 'space-born' show no interest in the colony's Earth-related goals. Instead they are drawn to the zero-gravity axis region, the most authentically non-terrestrial environment, where they develop new ways of moving, new forms of art. '[For the children] the Colony … is the only *real* world, the only one that matters … As for us [adults], we've served our purpose. We were just the transition team' (ibid., 348–350). Perhaps this is predictive of a problem for real-world colonies. Without careful social engineering and education, there seems no a priori reason why 'space-born' children should care remotely about a world they have never visited, or about goals devised by their parents long before they were born.

2.3.2 *The Space Enthusiasts*

With time, O'Neill's proposals inspired much more extensively developed visions of the 'high frontier', many of them quite utopian. From 1989 American author Allen Steele., in the early novels of his 'Near Space' future history sequence (1989, 1990), seized on the basic O'Neill plan and used it to spin dreams of blue-collar workers in space. While these books are ostensibly gritty and realistic, they are at the same time extraordinarily romantic—and are heavily influenced by similar works by Heinlein several decades earlier (compilation 1977). *Orbital Decay* (1989) is a projection from the then present in which, by the late 1990s, the major corporations have moved into space activities, notably Skycorp. Set in the year 2016, the drama is centred on Skycorp's wheel in space, the Olympus Station, known as 'Skycan' by the workers aboard. Nearby is the zero-gravity facility Vulcan Station, used to construct SPS satellites from lunar aluminium (ibid., 80). And under cover of 'Meteorology' studies, national security operatives are constructing a 'Big Ear', a covert facility capable of monitoring telephone and other conversations anywhere on the planet.

Steele deliberately contrasts this working environment with the 1950s von Braun visions, with their 'spit-and-polish Air Force types going around saluting and eating food capsules' (ibid., 33). Like an oil rig, the purpose of the enterprise is to extract energy from an inhospitable environment, and the workers fit the situation: 'These guys are mainly blue-collar, salt-of-the-earth, hard-hat types, with a wild-ass streak ... They don't want to hear discourses about a manifest destiny among the stars, they want to make a bundle at a high-risk profession and get home alive' (ibid., 210). However, Steele has his workers rise up against what they see as the anti-democratic activities of the Big Ear project; in the end they see themselves as pioneers in the American tradition, and it is 'the right of pioneers to decide what happens on the frontier' (ibid., 316).

The sequel, *Clarke County, Space* (1990), set a generation on in 2049, is about a more fully committed space colony. Hosting some 8000 people the eponymous colony is centred on the 'biosphere', a rotating sphere of radius ~ 110 m (ibid., 26). The conflict concerns the destiny of the habitat. Its inhabitants see it as a seed bed for the human expansion into space; it aims to become self-sufficient, it hosts agricultural experiments (ibid., 52), and there are dreams of spawning more colonies off in the asteroid belt. On the other hand in the here and now it is still a 'company town' (ibid., 53) and its corporate controllers, seeking a quick return on their investment, use it as a tourist resort. Again the frontier spirit prevails, and a movement begins for the colony to declare its independence: 'This colony—this community—will not be bought-and-paid-for by a bunch of corporate greedheads who want to turn it into a tourist trap.' (ibid., 55).

But for some writers O'Neill's vision was always ambiguous. Set somewhat further in the future, Katherine MacLean's 'The Gambling Girl and the Sinful Hell', a story in the generally positive Pournelle anthology (1979), is a tall story of a family homesteading the asteroids in a one-family spacecraft of the kind O'Neill advocates: 'Abe was getting too big for the home barrel ...' (ibid., 267). This folksy story of a widowed mother and her kids sharing their 'barrel' with chickens and piglets may echo fantasies of little houses on the prairie. But to many readers the confinement and isolation the children endure will seem stark: 'The girl was staring around at a circle of faces ... We'd hardly seen anyone new except Sam and MacPherson whose orbit was almost the same as ours ...' (ibid., 273). Isolation and a dependence on communal systems for the basics of survival could of course make small or large colonies naturally tyrannous environments (Cockell 2013), in direct opposition to O'Neill's dreams of freedom and progress.

Thus even the most positive of stories about O'Neill colonisation could contains seeds of doubt. And with time more critical fictions would be written.

2.4 Dystopias in Space

2.4.1 Economic Doubts

Through the 1980s the O'Neill model was closely inspected in fictional works and beyond, and doubts were formulated, objections raised. For example, against a background of a reduction in energy costs after the oil crises of the 1970s, the economic model for the space islands' proposed development based on SPS looked less promising.

Trojan Orbit by Reynolds and Ing (1985) is an entertainingly searing critique in fictional form of the O'Neill vision. In the (then) near future, while the Soviets patiently build a modular station of the Mir-ISS type, the west has invested in the O'Neill dream, with 'Island One' having been established at L5, whose inhabitants are intended to be building SPS plants and further colonies. However, the authors argue, the practicalities of the project have simply not been thought through. They quote a paper of O'Neill's in *Futurist*: 'The first space community would house 10,000 people; 4,000 would be employed building additional colonies, while 6000 would be producing satellite solar power stations'. 'Wizard, but who was supposed to be running the island? Who was going to be keeping the hydroponic farms going, regulating the air and water …? Who was going to be teaching the kids? Who was going to be taking care of the hospitals?' (O'Neill 1976b, 129). It ultimately emerges that the colony is a huge racket, controlled by organised-crime families in order to siphon off the billions of dollars' worth of investment in the station. The book is dated and lurid, but perhaps it should be required reading for all O'Neill advocates.

Meanwhile, aside from the economics, how would it be to live in such habitats?

2.4.2 Cages in Space

Space colonies, floating in the vacuum, may paradoxically feel like burrows in the ground. In addition to metres-thick layers of moon rock to provide radiation shielding, plants grown in space would need windows of lunar glass ~ 10 cm inches thick to protect them from raw, unfiltered sunlight. The inhabitants would not even be able to see out, to see that they were in space. Such habitats could seem very unwelcoming places, and this was reflected in fiction. One ghastly glimpse of the result of long-term exposure of workers to microgravity is Kelly's story 'Breakaway, Backdown' (1996): 'Her muscles have atrophied so her papery skin looks as if it's been sprayed onto her bones … "I've got 40 % bone rot … and I mass 38 kilos … This is how space makes us over.'

A brand new space habitat would no doubt be an attractive destination. But what happens when the technology grows old and break down? In Sterling's *Schismatrix* (1985), a dramatic vision of a posthuman future in the solar system and beyond, the

Concatenate is a federation of O'Neill-type colonies orbiting the moon, habitats built in the twenty-first century but by the book's opening in the twenty-third century historical relics. Like modern-day Detroit, the 'Mare Tranquillitatis People's Circumlunar Zaibatsu' is a city in space that has become bankrupt. Entering, the protagonist 'could stare the length of the Zaibatsu, through five long kilometres of gloomy, stinking air …' (ibid., 11) Internal society has broken down, with people living in shacks built from ruined factories and sealed against the disease-laden air (ibid., 22). The ghastly truth is that the inhabitants of this orbital slum have nowhere to go, no chance of economic recovery, no prospect of salvation from their plight.

Another troubling aspect of space habitats is their inherent fragility. O'Neill dismisses the dangers of terrorism to space habitats, thanks to the possibility of screening at limited access facilities, and, so O'Neill claimed, the difficulty of an individual doing large scale damage to a habitat (1976a, 111). But Sterling (1985) argues that living in such fragile habitats would condition the psychology of the populace: '*Worlds could burst* … Outside those locks loomed utterly pitiless darkness …There was no true safety … There were a hundred ways to kill a world: fire, explosion, poison, sabotage … The power of destruction was in the hands of anyone and everyone … The spectre of destruction had shaped the moral paradigm of every world and every ideology' (Sterling 1985, 64).

In Gibson's *Neuromancer* (1984) L5 habitats, part of the furniture of a heavily corporate future, are presented entirely negatively: 'Archipelago. The islands. Torus, spindle, cluster. Human DNA spreading out from gravity's steep well like an oil slick' (ibid., 125). Even the builders of Freeside, a massive cylindrical habitat (ibid., 132) which, with its hotels, brothels and casinos, dominates an archipelago of settlements (ibid., 149), have turned inwards, creating a colony hidden within the habitat which is compared in horrific terms to a wasps' nest (ibid., 204).

This peculiar introversion, this burrowing inward, is a common feature even of utopian visions of space habitats. In the superficially attractive cylinder-world glimpsed in the movie *Interstellar* (2014), it is impossible to see out into space, and the architecture is that of the past, of an idealised American small town folded on itself. It is almost as if the characters are not in space at all. This sense of a retreat inwards and to the past can feel at odds with the generally progressive, future-oriented nature of much SF discourse.

Meanwhile, other authors have depicted O'Neill habitats not as shelters for workers but as castles in the sky for elites.

2.4.3 A Celestial Elite

The idea of space habitats housing a *benevolent* elite is featured for example in Sagan's *Contact* (1985). The attraction of space for the elderly wealthy is the suggestion of longevity in zero gravity conditions: there is 'the faintest aroma of immortality' (ibid., 281). By the year 2000 there are 'rudimentary retirement hotels' in Earth orbit. There are qualms: 'It was foolhardy, they said, to permit an elite class to emigrate to space, with the masses left back on Earth—a planet in effect given over to absentee landlords'. But Sagan takes an optimistic view of the effect of space on its elite colonists: 'Hardly anyone anticipated the principal outcome, the transfer of a vivid planetary perspective to those who would do the most good' (ibid., 282). Indeed it is a consortium of the orbital wealthy who lead the final construction of the alien 'Machine' that takes Ellie Arroway to the stars. Speculation on the medical benefits of space habitation had already dated back decades; see for example Clarke (1968, 151), who had outlined possible advantages for serious burns victims, post-operative therapy, and the 'possibility—wildly speculative … that the expectation of life may be increased when the wear and tear of gravity is removed.'

Meanwhile the idea that a wealthy elite in space habitats may *not* necessarily prove to be benevolent has been explored since some of the earliest fictional reactions to O'Neill's pioneering studies. In particular, it is surely a weakness of the O'Neill blueprint that the planet's vital energy supply could be easily controlled by a handful of people in space. In Ben Bova's *Colony* (1978), in the year 2008 Island One is an O'Neill cylinder, 'landscaped, filled with air, an engineered paradise that housed an elite few of very rich people—while billions lived in misery on the tired, crowded old Earth' (ibid., 10). The habitat is the hub of a solar power industry. Five super-wealthy individuals known as the Board are controlling access to space power; they seek to destroy a World Government which, by trying to force them to use their profits to alleviate social problems, they see as an obstacle to their own ambitions. Ultimately Island One, and a private second cylinder, will be the final refuge of the super-rich, while Earth burns (ibid., 107).

In Joe Haldeman's *Worlds* novels (1981–1992), in the 2080s 21 space habitats, 'Worlds', orbit an overcrowded Earth. The largest is New New York, with a quarter of a million inhabitants. While politically independent, New New York is economically in debt to the US after cheap fusion ended the economic justification for SPS, and it depends on organic materials from Earth—but when lodes of such material are found on the moon, the prospects of the Worlds are transformed. However this initiates tensions with Earth. The crisis comes when New New York, in a show of force, cuts the power from its SPS stations to the US. After a devastating war the Worlds become refuges of civilisation, orbiting a ruined Earth. The 2013 movie *Elysium* (dir. N. Blomkamp) portrays a similarly bleak view of elitism in a space colony. Director Blomkamp was inspired by a *National Geographic* report on the 1970s Stanford Torus design (Johnson 1977) to imagine a kind of gated community in space, ethereally beautiful; the half-million citizens of Elysium, having fled to the sky, ruthlessly exploit an Earth ruined by environmental collapse and over-exploitation. At least one veteran of O'Neill's work (Brody 2013) objected to the subversion of utopian studies from 1970s California into a twenty-first-century portrayal of an instrument of oppression.

2.5 New Social Orders: Fragmented Cultures and Limits to Growth

2.5.1 The Fragmentation of Mankind

As noted above, one distinctive feature of O'Neill's scheme is that, despite his famous designs for large space habitats, he sees a long term future in which freedom for mankind is secured through its scattering into a series of much *smaller* communities. More generally, O'Neill argues (1976a, 17) for any technological improvement being beneficial only 'if it reduces rather than increases the concentration of power and control … if [such improvements] *tend to reduce the size of cities, industries and economic systems to small size*, so that bureaucracies become less important and direct human contact becomes more easy and effective' (my italics). The 'evils of bigness' include 'high crime rates … social alienation, and political corruption' (ibid., 39). And human communities need room to experiment. Since we have yet to have found an ideal government form, 'what chance for rare, talented individuals to create their own small world and family, as was so easy a century ago in our America as it expanded into a new frontier?' (ibid., 40).

In addition O'Neill sees growth as a buffer to freedom and happiness. O'Neill argued that human freedom could be assured by giving people the ability and the room to move and build a new society for themselves. He argues against imposed limits of all kinds: 'The freedom to have as many children as a family wants is by no means as important as the freedoms of speech, communications, travel, choice of employment, and the right to an education, but it is hard to abrogate one freedom without compromising others' (1976a, 246). In O'Neill's model of the future, it may seem that the evolution of human society is driven by irreparable flaws in our own nature. Our inability to build stable large communities must lead to the fragmentation of society, and our inability to control our population numbers must lead to endless fissioning, movement and growth.

But are there plausible, and desirable, alternatives?

2.5.2 Melting Pots

There are in fact technical arguments in favour of large habitats rather than small, such as given in Fogg's discussion of contained biospheres (1995, 48ff). While the functioning of biospheres is imperfectly understood, Fogg argues that it may be impossible to scale down Earth's biosphere by many orders of magnitude (five orders down from Earth to an O'Neill cylinder) and expect it to maintain all its functions adequately. And the smaller the size of container, the more conscious intervention is likely to be required maintain the habitat.

In addition there may be scientific or other reasons why large habitats could be desirable. For example, could space habitats serve as wildlife refuges? In Robinson's *2312* (2012), set in the twenty-fourth century as the title suggests, mobile habitats called 'terraria' (ibid., 36–40) are typically hollowed-out asteroids comparable in size to O'Neill's Island Three. There are nineteen thousand terraria, some given over to farming, others used as reserves for species threatened on a post-climate-change Earth. There are even ecologies containing creatures extinct but restored, such as a terrarium called *Pleistocene* containing Ice Age flora and fauna (ibid., 59). This idea dates back to suggestions by O'Neill himself that space islands could be used as wilderness refuges (1976a, 253). But this too was predated by the wistful vision of the movie *Silent Running* (dir. D. Trumbull, 1971) which showed domed forest reserves held in orbit around Saturn, with the ultimate intention being to 'refoliate' an Earth that seems to have become a bland, nature-free utopia.

Note however that Robinson's terraria are not very large in terms of the space needed by wildlife in nature. A wolf pack, consisting of ~ 10 animals, may have a territory of 35 km^2 (Jędrzejewski et al. 2007). A Robinson terrarium with an inner surface area of ~ 160 km^2 would have room for only ~ 4 packs, or ~ 40 individual animals, a small population in terms of genetic diversity and the salvation of a species. Even an O'Neill colony is probably too small to contain wilderness.

As regarding social issues, given prior examples on Earth, even if a peaceful partitioning of communities is achieved it may not always be a happy solution. How to decide, among the descendants of the pioneers who built a habitat, who should stay and who should go? And what may look like a healthy parting of the ways to one group might look like cleansing (ethnic, religious, ideological) to another.

One American voice to provide a counter-argument against the fragmentation of mankind in space was Isaac Asimov, in his novel *Nemesis* (1989). In the 23rd century the solar system is divided between an overcrowded Earth and a sky full of 'Settlements'. The Settlements stand aloof from Earth, which they regard as an 'unliveable slum' (ibid., 47)—and also from each other, if only for the fear of infection from diseases bred in separate, isolated biospheres: 'Commerce is being throttled for fear of picking up someone else's strains of parasites or pathogens' (ibid., 29).

Further, the Settlements themselves are portrayed as unhealthily cleansed socially: 'On any Settlement, all are alike, or, if there is some admixture to begin with, those who are well outnumbered feel ill-at-ease, or are made to feel ill-at-ease, and shift to another Settlement where they are not outnumbered …' (ibid., 130). This is a rejection of a tradition of relative tolerance which perforce has had to evolve on Earth. 'We're talking about Earth's long struggle to find a way to live together, all cultures, all appearances. It isn't perfect yet, but compared to how it was even a century ago, and it's heaven. Then when we get a chance to move into space, we shuck it all off and move right back into the Dark Ages' (ibid., 156–157).

There have been other wistful depictions of large space habitats as places of peaceful encounters: 'It was the dawn of the third age of mankind, 10 years after the Earth-Minbari War. The Babylon Project was a dream given form. Its goal: to prevent another war by creating a place where humans and aliens could work out their differences peacefully … Humans and aliens wrapped in two million, five hundred thousand tons of spinning metal, all alone in the night … The year is 2258. The name of the place is Babylon 5' (opening narration, season 1). The deep space station Babylon 5, star of the eponymous TV series (Babylonian Productions, 1994–1998), is a large O'Neill cylinder located in orbit around a planet of Epsilon Eridani, and capable of supporting 250,000 human-scale entities in a variety of gravity regimes. In a crowded and conflict-filled Galaxy, the station was established at a contact point of five major interstellar powers, and became a junction of inter-species diplomacy.

In the TV series *Star Trek: Deep Space Nine* (Paramount 1993–1999), meanwhile, the eponymous space station, a glorious, Gothic wheel of immense proportions, was built in orbit around the planet Bajor by the occupying Cardassians. When the occupation was lifted the Bajoran government invited the United Federation of Planets to jointly administer the station, and with a nearby wormhole offering access to the 'Gamma Quadrant', a remote part of the Galaxy, the space station becomes a multicultural centre for interstellar exploration, trade, politics. Even in the movie *2001: A Space Odyssey* (1968), a space station meant as a mere waystation was a place where Cold-War American and Soviet scientists could at least meet over a drink.

2.5.3 Freedom Within a Habitat

As we have seen O'Neill argues that the ability to escape a habitat is a necessary buffer to human freedom. But is a free society possible without this safety valve? Of course even the largest single habitat must have limits to population growth—and there may be other unexpected constraints on the inhabitants' freedoms. Grant (1984), in the context of world ship designs, studied the stability of populations in such habitats under various regimes. The results of his computer-modelling are complex, but since in the long term a *fall* of population, causing a loss of capability and genetic variability, is as damaging as a resource-depleting population excess, it may be that in some circumstances inhabitants would be compelled to *have* children.

More generally, is freedom possible without growth? A *static* society, of the kind described in the Club of Rome's 'Limits to Growth' report (Forrester 1971), is economically at least a feasible solution to the conundrum of survival in a space colony. O'Neill (1976a, 27) himself points out that steady-state societies are possible, and cites the pre-Conquest Inca empire as an example—but a negative one: 'at his death [an Inca peasant] … left a world almost exactly the same as the one he was born into'. In his view, such a society is 'rigidly structured, dictatorial'; 'almost any static society is forced in self-defence to suppress new ideas'. In the author's own fiction the Inca-dominated habitat in *Ultima* (2014) is a static society but an autocratic one; excess population is creamed off for labour on extraterrestrial mines, the military and other activities.

Nevertheless, in a world that seems even more tightly constrained by resource limits than those faced by the authors of the 'Limits to Growth' studies of the 1970s, there are modern studies in how prosperity, including spiritual growth, could be achieved without endless material and economic growth (Jackson 2009). Possibly by the time we inhabit large colonies in space we will necessarily have learned how to live within steady-state societies on Earth without compromising our essential human freedoms; with new goals and new motivations, such ways of living might be readily transferred to life in space habitats.

But, even if individuals can escape from habitats to go homesteading in the solar system, there are still more fundamental limits to growth.

2.5.4 A Finite Solar System

O'Neill argues that modest growth 'will encourage the extension rather than the curtailment of freedom … I've argued that a growth rate about a tenth as large as the present explosive increase would make the difference between stasis and change …' (1976a, 247). Even at low rates, growth requires room. O'Neill imagined individuals leaving large habitats and homesteading new terrains, a movement that would naturally lead to a solar system full of colonies. We glimpse swarms of diverse habitat-based communities in Gibson's *Neuromancer* (1984), and Robinson's *2312* (2012) shows a solar system full of inhabited asteroids. But there are limits to growth even in the solar system. O'Neill himself speculates (1976a, 247) that the resources of the asteroids, equivalent to the surface area of 3000 Earths, might fill up, even at the modest growth rates he predicts, in a few thousand years; exploiting the more remote resources of the solar system might allow expansion for several more thousand years.

And what then? Is our destiny, driven by the imperative to growth, to move beyond the solar system?

2.6 An Arena of Ultimate Dreams

Since the works of Tsiolkovsky (1920) and Bernal (1929) some thinkers have speculated on the largest scales about the ultimate destiny for humanity and human civilisation in space. The author's own 'Open Loops' (2000) is an account of a future colonisation of interstellar space by post-humans fully adapted to the zero gravity conditions of asteroid habitats. Ten thousand years after its first colonisation, asteroid Ra-Shalom is a ball of liquid containing fish-like post-humans subsisting in an ecology essentially shared with blue-green algae. After a million years, the nearby stars glow green, surrounded by clouds of such habitats.

Perhaps the ultimate fictional realisation of these most spiritually expansive of space-colony dreams is Zebrowski's *Macrolife* (1979). Zebrowski references O'Neill, though his direct inspiration (afterword) was the work of Cole (1961). In the year 2021, mankind's first large space habitat is Asterome (Chapter 6), a 'mobile utopid' located at L5, a hollowed-out asteroid some ten miles long and five miles across. Hosting 100,000 people, it has become a centre of industry, research and colonisation. However a disaster overwhelms the Earth and Asterome flees to the stars. By the year 3000 new 'macroworlds' have been produced at such stars as Alpha Centauri and Procyon. Asterome itself has some properties of a larger life form, 'macrolife'; it is able to 'reproduce' by shedding outer layers, to leave a hollow shell into which a raw asteroid is taken for reworking. It now emerges that macroworlds are a convergent goal for many forms of life: 'Macrolife was the brain and nervous system of something being born all over the Galaxy' (chapter 25). A staggering coda to all this is set a hundred billion years hence, as the universe faces a termination through a Big Crunch. Some of the macroworlds manage to survive the terminal singularity, and encounter what appear to be relics of still earlier cycles.

All this is a long way from O'Neill islands at L5, huddled in their cloaks of moon rock, earning money from clunky solar-power stations. But perhaps a space advocate would argue that such visions express the ultimate ethical choice concerning space colonisation: to ensure the survival of the human species into the very far future, or not. On the other hand, perhaps O'Neill is wrong about the fundamental imperatives behind mankind's growth; perhaps we will after all learn to live within our means. As SF writer Brian Aldiss once remarked from the audience in response to a speech of the author's on expansion in space, 'But we've heard all this before! If only we could get along with each other, we wouldn't have to go to all the trouble of conquering the Galaxy!' (Novacon 23, 1993).

2.7 Conclusions

'The glass sunlight panels were coated with filth … A cadre of lumpy robots were scraping and mopping the fretted glass … Lindsay realised suddenly that they were human beings in suits and gas masks' (Sterling 1985, 13).

Writers of science fiction have responded imaginatively to the proposals for space colonisation developed by O'Neill and co-workers in the 1970s. Post-publication response to O'Neill's utopian vision in SF has been positive from some writers who welcomed to the idea of a new frontier. But others were critical, foreseeing such drawbacks as the destruction of the health of space workers, the grimness of life in a failing habitat, coercion by corrupt elites, and the unhealthy fragmentation of mankind.

In the forty years since O'Neill's first publication, continuing fictional explorations constitute a bank of thought experiments on how O'Neill's ideas might play out in reality and their impact on humanity. As to the predictive accuracy of these tales, only time will tell.

References

Asimov, I. (1989). *Nemesis*. New York: Doubleday (page numbers from the 1990 Bantam edition).

Banks, I. M. (2004). *The Algebraist*. London: Orbit.

Baxter, S. (2000). "Open Loops", in Benford and Zebrowski, 2000.

Baxter, S. (2014). *Ultima*. London: Gollancz.

Bernal, J.D. (1929). *The World, the Flesh and the Devil*. London: Jonathan Cape (page numbers from the 2010 Prism Key Press edition).

Benford, G., & Zebrowski, G. (2000). *Skylife*. New York: Harcourt.

Bova, B. (1978). *Colony*. New York: Pocket (page numbers from the 1989 Mandarin edition).

Brody, D. (2013). 'Not "Elysium" but Better "Ringworld" Settlements Could Return Our Future to its Past'. www.space.com/22326-elysium-movie-space-colonies-future.html. Accessed August 31, 2013.

Clarke, A. C. (1968). *The promise of space*. London: Hodder and Stoughton (page numbers from the 1970 Pelican edition).

Cockell, C. S. (2013). *Extraterrestrial liberty: An enquiry into the nature and causes of tyrannical government beyond the Earth*. Edinburgh: Shoving Leopard.

Cole, D. (1961). *Social and political implications of the ultimate human society*. New York: General Electric, Missile and Space Vehicle Department.

Fogg, M. (1995). *Terraforming*. New York: Society of Automotive Engineers Inc.

Forrester, J. W. (1971). *World dynamics*. Cambridge, Mass: Wright-Allen Press.

Gibson, W. (1984). *Neuromancer*. New York: Ace (page numbers from the 1995 Voyager edition).

Grant, T. J. (1984). The population stability of isolated world ships and world ship fleets. *Journal of the British Interplanetary Society, 37*, 267–284.

Haldeman, J. (1981). *Worlds*. New York: Pocket Books.

Haldeman, J. (1983). *Worlds apart*. New York: Ace Books.

Haldeman, J. (1992). *Worlds enough and time*. New York: Pocket Books.

Heinlein, R. A. (1977). *The past through tomorrow*. New York: Berkley Medallion Books.

Jackson, T. (2009). *Prosperity without growth? The transition to a sustainable economy*. Report for the Sustainable Development Commission. Summary http://research-repository.st-andrews.ac.uk/bitstream/10023/2165/1/sdc-2009-pwg-summary.pdf. Accessed November 18, 2014.

Jędrzejewski, W., Schmidt, K., Theuerkauf, J., Jędrzejewska, B., & Kowalczyk, R. (2007). Territory size of wolves Canis lupus: Linking local (Białowieża Primeval Forest, Poland) and Holarctic-scale patterns. *Ecography, 30*, 66–76.

Johnson, R. D., & Holbrow, C. (Eds.). (1977). "Space Settlements: A Design Study", NASA SP-413.

Kelly, J. P. (1996). Breakaway, Backdown. *Asimov's Science Fiction*, June 1996.

O'Neill, G. K. (1976a). *The high Frontier*. New York: William Morrow (page numbers from the 1978 Corgi version of the second edition).

O'Neill, G. K. (1976b). Space colonies: The high Frontier. *The Futurist*, February 1976.

Pournelle, J. (1979–82). *The endless Frontier* (Vol. I, 1979; Vol. II, 1982). New York: Baen.

Reynolds, M., & Ing, D. (1985). *Trojan orbit*. New York: Baen.

Robinson, K. S. (2012). *2312*. London: Orbit.

Ryan, C. (Ed.). (1952). *Across the space Frontier*. New York: Viking Press.

Sagan, C. (1985). *Contact*. New York: Simon & Schuster.

Steele, A. (1989). *Orbital decay*. New York: Ace (page numbers from the 1990 Arrow edition).

Steele, A. (1990). *Clarke county, space*. New York: Ace (page numbers from the 1991 Arrow edition).

Sterling, B. (1985). *Schismatrix*. New York: Arbor House (page numbers from the 1996 Ace edition of *Schismatrix Plus*).

Tsiolkovsky, K. (1920). *Vne Zemli* (*Beyond the Planet Earth*) (K. Syers (1960), Trans.). New York: Pergamon Press.

Westfahl, G. (2009). *Islands in the Sky*. New York: The Borgo Press.
Williamson, J. (1931). 'A Prince of Space'. *Amazing Stories*, January 1931. New York: Teck Publications.
Zebrowski, G. (1979). *Macrolife*. New York: Harper & Row.

Chapter 3
Space Colonies and Their Critics

Gonzalo Munévar

3.1 Introduction

There may be no better example of a utopia based on outer space in the 1970s than Gerard O'Neill's proposal for building space colonies, as presented in his book *The High Frontier* (1976/1982). At the same time O'Neill's case also exemplifies the reaction of the "limits to growth" advocates, for their attack on his ideas was as swift as it was vehement. O'Neill drew inspiration from several space pioneers whom had envisioned boundless possibilities for humankind in the exploration and colonization of the cosmos. He was particularly impressed by the work of European thinkers such as Konstantin Tsiolkovsky, John Bernal, Hermann Oberth, Guido von Pirquet, Hermann Noordung, Wernher von Braun, and Krafft Ehricke. His ideas for "islands in the sky", for example, owe much to Bernal spheres, while his emphasis on finding solutions to our energy, pollution, and scarcity problems by building solar power satellites and mining the Moon and asteroids echoed Ehricke's "extraterrestrial imperative" to sustain the development of humanity by exploiting the resources of the solar system. His environmentalist critics believed instead that the very attempt to escape our limits by going into space was irresponsible day-dreaming. Wendell Berry, for instance, called him shallow and gullible, a thug. And Dennis Meadows urged us to solve our problems down here instead. Nevertheless, in spite of the bitterness of the controversy, both sides created the basis for cooperation decades later, making us dream again of the utopias once dreamed by O'Neill.

G. Munévar (✉)
Lawrence Technological University, Southfield, MI, USA
e-mail: gmunevar@ltu.edu

J.S.J. Schwartz and T. Milligan (eds.), *The Ethics of Space Exploration*,
Space and Society, DOI 10.1007/978-3-319-39827-3_3

3.2 The European Antecedents of Space Colonies

The utopian idea of large space settlements in space, not on the surface or the body of a planet, i.e., "space colonies", is rightfully associated with the name of the American physicist Gerard O'Neill. Nevertheless, as he himself acknowledges in his famous book *The High Frontier*, he was the heir of a long series of European space visionaries (O'Neill 1982, 2) both in the technical details and in the motivation for undertaking the enterprise. Indeed, it was the Russian space pioneer Konstantin Tsiolkovsky who first addressed both matters in his novel *Beyond the Planet Earth*, written in serialized form around 1900 and then published as a book in 1920.[1] Tsiolkovsky wrote of "mansion-conservatories" in geosynchronous orbit filled with men, women and children, in an environment awash with "a thousand more times solar energy than the Earth…it only remains to fill it with dwellers, greenhouses and people" (ibid, 61). Tsiolkovsky clearly explained artificial gravity, to be produced by having the colony rotate, and stressed the importance of having a large enough radius of rotation (to avoid coriolis forces in the bodies of the space dwellers). He also conceived of parabolic mirrors of "unlimited size", thanks to the lack of gravity, which could be used to supply large amounts of energy. This flight into space, O'Neill points out, was motivated by the same circumstances that prompted O'Neill himself to propose his own version of "mansion-conservatories": a terrestrial population that is beginning to feel the ecological limits of the planet. To the coming ecological crisis, Tsiolkovsky saw the same sort of solution that O'Neill would envision some 75 years later: the mineral resources of the asteroids in the context of the industrialization of the solar system through space colonies. As Tsiolkovsky wrote: "The high temperature, the chemical and thermal energy of the Sun's rays, not weakened by the atmosphere, make it possible to carry out all kinds of factory work, such as metal welding, recovering metals from ores, forging, casting, rolling, and so forth" (ibid, 62).

O'Neill pointed out that "if we were to excavate the land area of the Earth to a depth of half a mile, and to honeycomb the terrain to remove a tenth of all its total volume, we would obtain *only* 1 % of the materials contained in just the three largest asteroids" (ibid, 60); emphasis added). Thus both writers were motivated not only to provide a future for those who left the home planet but also to preserve the Earth for those who remained. O'Neill also envisioned producing abundant clean energy for the Earth by constructing in orbit gigantic solar collectors (the size of Manhattan Island) and safely beaming non-pulsated microwave energy down to

[1]O'Neill's references to Tsiolkowsky appear in (O'Neill 1982, 2) and (ibid, 60–62). Tsiolkowslky's more formal work on rocketry can be found in his *The Rocket into Cosmic Space*, Moscow, Naootchnoye Obozreniye (1903). K.K. Lasswitz also explored the idea of habitats in space about the same time. *On Two Planets*, Leipzig, 1897.

rectifying antennas ("rectennas") on the ground, where the energy would be transformed into electricity.[2]

Largely independent of Tsiolkovsky's bold vision, other European space pioneers, inspired by the German engineer Oberth's (1923) seminal work on rockets,[3] contributed ideas on space habitats that would later influence O'Neill. One of those pioneers was Guido von Pirquet, an Austrian scientist who calculated many of the preferential spacecraft trajectories to the planets (one actually taken by a Russian mission to Venus) and was a strong advocate of space stations.[4] Another Austrian, Hermann Noordung, conceived of a torus-shaped space station, a Wohnrad[5] (Figs. 3.1 and 3.2). Wernher von Braun later designed a similarly shaped space station for NASA (Fig. 3.3), which Stanley Kubrick used as a model in his movie *2001: A Space Odyssey*. Noordung's model was too small and his suggested rotation would have caused problems for the astronauts. He later imagined habitats thousands of meters across. Some of them could be seen as forerunners of O'Neill's cylindrical design of his Island Three, which would be presumably several kilometers at the base and over 30 km in length, housing millions of human beings (Fig. 3.4).

Of crucial importance in the conception of space colonies was the work of J.D. Bernal, an Irish scientist whose main area of expertise was X-ray crystallography, not space technology. In his (1929) space utopia he proposed a non-rotating, and thus non-gravitational, sphere 1.6 km in diameter.[6] Almost half a century later, O'Neill would find that shape ideal for colonies within a certain range of population, both in terms of the useable land and ratio to surface area per unit volume. His Island One is thus a *rotating* Bernal sphere 500 m in diameter, which would allow for comfortable artificial gravity at its equator. His Island Two was a little larger than the one Bernal had proposed, 1.9 km, and could be home to as many as 140,000 people.

Although O'Neill echoes many of the ideas of his predecessors in his book, he claims, with some justice, that "it would have been difficult before the year 1969 to make them a coherent picture without serious technical gaps" (O'Neill 1982, 2). Nevertheless, in forming such a coherent picture, he relied on the talents and dedication of many scientists and space engineers. One of the most prominent was another European, the German rocket propulsion engineer Krafft Ehricke, a friend whose originality and drive O'Neill much admired for they "can be seen in ideas related to almost every area of development in space" (ibid., 311). But apart from

---

[2]This marked a shift on O'Neill's thinking, due to the influence of Peter E. Glaser, who had developed the concept of solar power satellites earlier (1968).

[3]Herman Oberth's book was his rejected doctoral dissertation, *Die Rakete zu den Planetenräumen* (By Rocket into Planetary Space), later published by München, R. Oldenbourg, (1923).

[4]O'Neill reports to have been influenced by the articles von Pirquet published in Die Rokete, vol. II, 1928.

[5]Noordung (1929). He used a nom de plume. His real name was Herman Potočnik. He was born in Croatia, then part of Austria-Hungary.

[6]Bernal (1929).

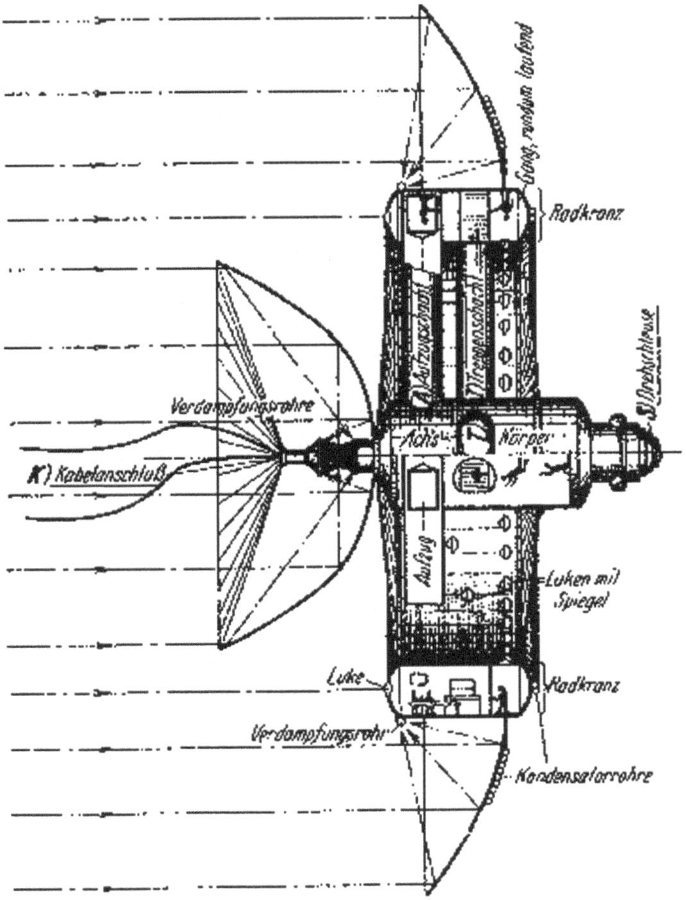

Fig. 3.1 Detailed Wohnrad diagram from Noordung's book (1929)

such technological contributions, O'Neill was also influenced by Ehricke's philosophical approach to space exploration, and particularly by his "Extraterrestrial Imperative". As Ehricke wrote:

> Our world is open to the cosmos and contains all the future and growth potential the human mind can envision … Such a concept is the realization that we can enhance the "supplies" to spaceship Earth, beyond energy, to include materials and information acquisition/transfer for the mainstream of human civilization … The Extraterrestrial Imperative is a manifestation of larger evolutionary cycles–an integral part of life's commitment to expansion and growth … This concept permits us to see beyond what seems to be an irreconcilable confrontation of man and environment …(Ehricke 1978).

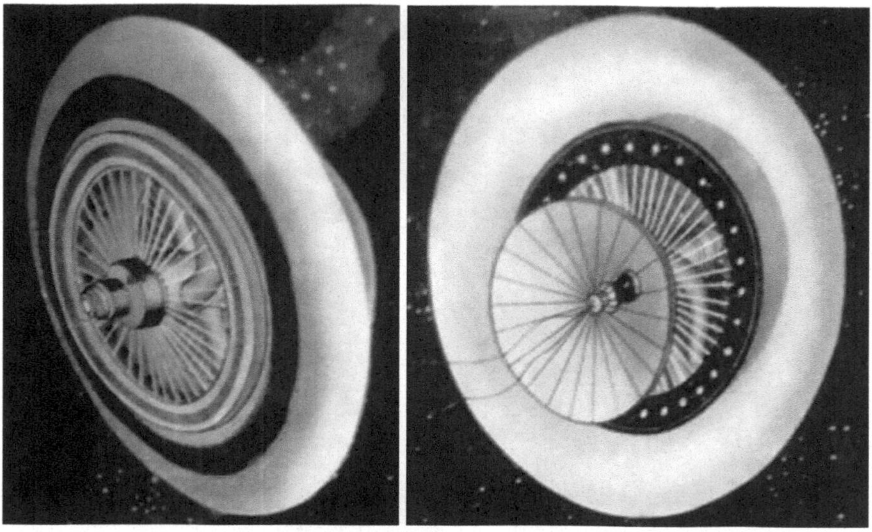

Fig. 3.2 Noordung's torus-shaped space station with parabolic mirror for concentrating sunlight. Courtesy of NASA

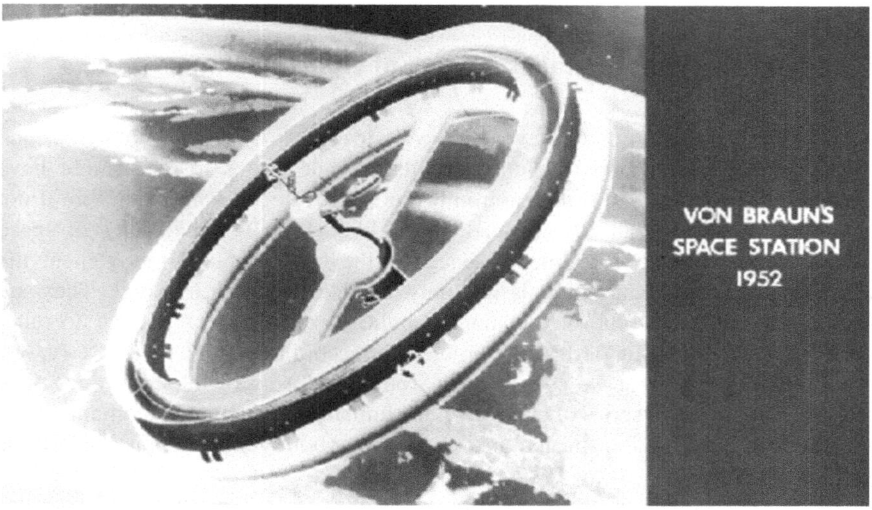

Fig. 3.3 One of von Braun's early space station designs. Courtesy of NASA

Fig. 3.4 A version of O'Neill's Island one, based on a Bernal sphere. Rings on top of sphere would be individual torus areas probably devoted to agriculture. Painting by Rick Guidice. Courtesy of NASA

3.3 The Environmentalist Critics of Space Colonies

Since O'Neill's vision was to prevent an ecological crisis by bringing materials and energy to our planet, and to reduce pollution while doing so, one might have imagined that the leaders of the environmentalist movement would hail him, if not as a hero, at least as an ally. Instead they vilified him. For the critics of space colonies, and especially for some influenced by the environmentalist movement, the very idea of space exploration is not only unwise, but also immoral. They are particularly harsh to some of the more grandiose proposals for going into outer space to solve pressing terrestrial problems. According to Wendell Berry, for example, the lesson that we should learn from the closing of the earthly frontiers "calls for an authentic series of changes in the human character and community that, if made, will afford us the spiritual resources to live both within our material means and with each other" (Berry 1977, 36).

Space exploration, he thinks, tries to outflank the lesson entirely. The space enthusiast—and here Berry has Gerard O'Neill in mind—ignores what is essentially a moral problem (i.e., the changing of human character and community) and offers technological solutions instead. The morality of the space enthusiast is thus both shallow and gullible, for he offers "a solution to moral problems that contemplates no moral change" (ibid.). Space exploration, to someone like Berry, could only be "a desperate attempt to revitalize the thug morality of the technological specialist, by which we blandly assume that we must do anything whatever that we can do" (ibid, 37). According to another critic, Dennis Meadows, "What is needed to solve these problems on earth is different values and institutions—a better attitude towards equity, a loss of the growth ethic.... I would rather work at the problems here" (Meadows 1977, 40).

At first sight Berry seems to beg the question. According to him, the closing of the earthly frontiers presents a moral problem to which only moral solutions are applicable. Gerard O'Neill and other space enthusiasts ignore the moral problem. Thus, Berry concludes, the space enthusiasts are not only doomed to failure but are also immoral (not just mistaken or unperceptive). But what O'Neill, Ehrlicke, Tsiolkovsky, and the others question is precisely whether all the frontiers have in fact closed. And certainly, if those frontiers haven't closed, we have no reason to believe that we face a moral problem. In assuming that the high frontier is not a genuine option, Berry heaps moral blame on the space enthusiasts while begging the issue in question. But perhaps there is a more sympathetic reading of Berry's position. What he may have in mind is that the experience of the (partial) closing of the earthly frontiers is enough to show that Western man's approach to nature is inherently unwise, and thus that its extension through space exploration is destined to fail. On what grounds should we trust O'Neill's grandiose plans for gigantic solar power satellites, let alone those for artificial worlds (his space colonies)? Surely projects of such magnitude cannot be made plausible by mere theoretical proposals. How can we be assured that no essential detail has been left out?[7] The most straightforward way to resolve this issue might be to demonstrate the feasibility of increasingly more complex stages of these projects. O'Neill would have been quite agreeable to this suggestion, but Berry and many other ideological critics would probably resist it. The reason for resisting it is that to undertake such demonstrations we first need a large commitment to space exploration, since the demonstrations require that we build and operate very large structures in space. But given the poor record of big technology, Berry would say, why should we extend it the benefit of the doubt on such a scale?

The environmentalist critics are thus not impressed by the suggestion that space exploration can help correct some of the excesses and mishaps of technological civilization. Nor are they impressed by the claim that space exploration enables us to appraise better how critical our environmental situation is. Such is the wrong approach to the problem. What we need to do is prevent the situation in the first place. Space is, then, a delusion, for it offers more growth and technology to stop the mess caused by growth and technology. Of course, the more we foul up the

---

[7]Some question, for example, the belief that in just a few years we could build an entire ecosystem from scratch, as would be required in one of O'Neill's space colonies. In addition to that, proponents of the exploitation of the resources of the solar system are often very optimistic about doubtful technologies; for instance, they frequently make references to self-replicating machines. I discuss the implausibility of such machines, also called "von Neumann machines", in Munévar (2012).

world, the more space will look like a necessity. But this is a false technological panacea. It is rather like a pain reliever that keeps the patient from having the operation that will save his life. As Wilson Clark puts it, "[O'Neill] speaks in terms of a 'first beachhead in space,' evoking the image of greener grass on yonder hill. Unfortunately, we have little time in which to prevent the elimination of the vegetation altogether" (Clark 1977, 38).

The urgency of the situation, as these ideological critics perceive it, makes unwarranted our engaging in any more technological detours. Western man's approach has brought the world to the edge of crisis by marrying technology to the mentality of growth. This ideological criticism touches the heart of space exploration insofar as science is supposed to provide the promissory note that underwrites that marriage in the first place. Once again, the attempt to satisfy scientific utopias—whether the building of space colonies or any others where "big science" is concerned—may be seen as a disturbance, an interference with nature. The emphasis on beneficial results is only a smoke screen: In the long run only a change of attitude can be beneficial. Anything not in harmony with nature is bound to make us fail. In the eyes of the ideological critics, space exploration amounts to a distraction at a time of crisis—the siren voice that calls us from the cosmos sings the tune of our doom.

The second consideration is that we better protect the Earth by nurturing respect toward it than by preparing to move onto another nest. For otherwise we imagine that spoiling our present nest is regrettable but not an insurmountable loss. Learning to live within the confines acceptable to our mother planet is a wiser policy because we already know that we can lead dignified and fruitful lives here. By contrast, learning to live in space is only a promise. Can we bet the future of mankind on it? The greatest gift we can make to posterity is a beautiful Earth and the strength of character to live in harmony with it. In other words, we accomplish more by preserving the natural balances that have been so accommodating to human beings in the past, and by restoring such balances where modern life has already disrupted them. The result of exercising greater moral responsibility toward the world is a better world.

Space exploration, on the other hand, presumably would continue the disruption of the natural balance. If technology has already caused a mess, why should we expect better? Moreover, space exploration would be worse than a necessary evil, for it is not an enterprise that we can engage in just once before returning to a more pastoral existence. As Berry says, in condemning the scientific mind, "(1) It would commit us to a policy of technological 'progress' as a perpetual bargaining against 'adverse effects'. (2) It would make us perpetually dependent on the 'scientific' foretelling and control of such effects—something that never has worked adequately, and that there is no good reason to believe ever will work adequately" (Berry 1977, 83). Why could it not work? Because "when you overthrow the healthful balance of the relationships within a system—biological, political, or otherwise—you start a ramifying sequence of problems… that is not subject to prediction, and that can be controlled only by the restoration of balance" (ibid.). Berry's warning is that "if we elect to live by such disruptions then we must resign ourselves to a life of desperate (and risky) solutions: the alternation of crisis and 'breakthrough' described by E.F. Schumacher" (ibid.).

The first thing that deserves comment is this matter of disruption and restoration of balance. A very early and rather important disruption of natural balance took place when life was born and changed the chemistry of the planet. Another crucial and massive disruption of balance came when the oxygen liberated by life "poisoned" the atmosphere and the oceans. And this was followed by the adaptation of life to oxygen, with the subsequent destruction of the cozy arrangements between early life and the environment. Disruptions of similar magnitude were brought about by the appearance of complex organisms, by the formation of an ozone layer, which made the land available to life—would it have been better for life to stay in the oceans?—and then by the return of vertebrates to the water, which led to whales and dolphins. Ever since, the evolution of life has created new forms that have remade the environment anew, destroying the very memory of whatever balance had been struck previously, and leaving at best a few scattered fossils of what the Berrys of the time would have insisted on preserving.

The fact of the matter is that life has often created new opportunities for itself, unwittingly no doubt, and has always changed the balance between its different forms—most of which are now extinct. The biota of the planet has remade itself many times over. The natural balance assumed by the critics is merely a fiction, a temporary arrangement that would change even if there were no human beings around to mess things up. And surely life does not exhaust the range of natural causes that have brought about massive disruptions of balance. Do volcanic eruptions, droughts, and asteroids always make for small reversible changes? What may we say, incidentally, of the galactic disruption that forced the collapse of the pre-solar cloud into a planetary system? Of the earlier obliteration of what may have looked like states of cosmic equilibrium, and thus of natural balance? Which balance is it that we are morally obliged to restore?

Clearly humans are not the only creatures that transform their environment and interfere with it. R.C. Lewontin, S. Rose, and L.J. Kamin point out that:

> …all living beings both destroy and create the resources of their own continued life. As plants grow, their roots alter the soil chemically and physically. The growth of white pines creates an environment that makes it impossible for a new generation of pine seedlings to grow up, so hardwoods replace them. Animals consume the available food and foul the land and water with their excreta. But some plants fix nitrogen, providing their own resources; people farm; and beavers build dams to create their own habitat (Lewontin et al.1984).

The issue is not, then, one of disrupting balance and interfering with the environment. Perfect balance can be found only right before the birth of the universe and perhaps right after its death. Even then we do not really know. And to avoid interfering with nature would be out of character for living things, while impossible to achieve anyway. What O'Neill offered was indeed a way to minimize the impact of mankind upon our planet.

Eventually environmentalists came to that realization, at least to some extent, and have found common ground with the supporters of exploration. Thus we now find organizations such as The Earth and Space Foundation, of which Charles Cockell is the Chair, devoted to bridging environmental concerns and the exploration of space (Cockell 2006). And even more recent work by James S. J. Schwartz. (2011) defends the notion that the two movements can go hand in hand. That compromise may become even easier as new technologies make some of O'Neill's ideas seem less grandiose or far-fetched. For example, smaller (football-field size), and cheaper designs of solar power satellites are getting a good deal of attention and several demonstration projects have been proposed. These new developments suggest, incidentally, that the suspicions of the early environmentalist critics were not entirely misguided. For example, in a NASA-NIAC report (Mankins 2012) proposing a practical solar power satellite (SPS) we read:

> ...early SPS architectures were technically complex and unlikely to be economically viable...
>
> These initial SPS approaches suffered from a number of significant technical and programmatic challenges, including:
>
> (1) Low technology maturity;
> (2) Excessive weight, due in part to huge, high-voltage power management and distribution (PMAD) (up to 7000 MW at >10 kV across a gimbaled interface);
> (3) Projected development costs for a monolithic platform more than 20 times larger than the International Space Station;
> (4) The up-front expense of the required fleet of heavy-lift reusable launch vehicles (RLVs); for example two-stage-orbit (TSTO vehicles with payload requirements of up to 250 mT.... (ibid., 12).

As our new confidence in building large structures in space increases, as shown in this and other reports, solar power from space may still come to be seen as a reasonable alternative to traditional power plants for generating electrical energy.[8] And they will spur the building of even larger structures, perhaps even a version of Island One, as we will see in the next section.

3.4 The Future of a Utopia

O'Neill's optimism led him to hope that there might be people working and living in space as early as the 1990s, but surely by the early part of the twenty-first century. Nothing of the sort has happened nor seems likely to happen in the next two decades. Although his space colonies would have encountered far greater technological difficulties than he had imagined, eventually those could have been overcome. Nor was the opposition by environmentalists decisive. His utopia did not come to pass when he envisioned it simply because it depended on the Space Shuttle as a reliable and inexpensive means of transportation into low Earth orbit, the first step in setting up a mining outpost on the Moon. But the Space Shuttle turned out to be extremely unreliable and the most expensive system ever devised

---

[8]Even the U.S. Military is beginning to take seriously the possibility of solar power satellites. And the Space Frontier Foundation sponsors a website for an open discussion of this technology: http://spacesolarpower.wordpress.com/.

for placing materials in orbit. Eventually, given the commitment by NASA and other space agencies, this difficulty shall be overcome. In that expectation, the commitment by O'Neill, and by the visionaries who preceded him, to the resources of space is beginning to seem very attractive once again. In particular, the idea of abundant renewable energy from space solar power satellites is being given serious attention by several space agencies around the world.

For instance, Japan's 2013 Basic Plan on Space Policy calls for solar power satellites as one of the three main research programs for the utilization of space. Accordingly, the Japan Aerospace Exploration Agency (JAXA) is entertaining both standard microwave and laser transmission of energy from satellites (Goto et al. 2014). The microwave version, which continues Glaser's original idea, is advantageous in that "The appropriate wavelength microwave beam is unaffected by clouds or fog" (ibid.). It may also be supplied pretty much continuously. Its main disadvantage is the size of the satellite and the receiving platform (rectenna). A proposed artificial island/rectenna off the coast of Tokyo would be 3 km long. The main advantage of a laser SPS is that "The spacecraft and ground facilities are made compact, and a larger space system can be constructed by assembling small independent [laser SPSs]" (ibid.). It has one disadvantage in that "[t]he intense laser beam can damage human eyes", and another in that "[t]he laser beam is significantly affected by atmospheric disturbances and the weather" (ibid.). In spite of these disadvantages, JAXA is actively conducting research and development on laser SPS, thus summarized by Goto et al.:

> The first step of the L-SSPS demonstration is the 500 m horizontal laser energy transmission experiments, which were conducted in 2013 and technological data were obtained. The second step will be the 100–200 m vertical laser energy transmission experiments, which will be conducted in 2015. The orbital laser energy transmission experiments, from the ISS (International Space Station)—JEM (Japanese Experiment Module, Kibo) to the ground, are expected to constitute the third step. (ibid.)

A seemingly stronger commitment is being made by Japan to the development of a microwave SPS. According to Sasaki (2014), JAXA "now has a technology road map that suggests a series of ground and orbital demonstrations leading to the development in the 2030s of a 1 GW commercial system—about the same output as a typical nuclear power plant" (ibid.). If built along the lines favored by Glaser and O'Neill, the result would be a "structure weighing more than 10,000 metric tons and measuring several kilometers across" (ibid.).

A novel alternative now pursued by JAXA, as Sasaki explains, is to employ two huge free flying reflective mirrors "positioned so that between the two of them, they would direct light onto two photovoltaic panels 24 h a day" (ibid.). It is hoped that additional research on light structural materials, and on very high voltage cables (to carry the energy from the photovoltaic units to the transmitter) will increase efficiency while bringing costs down. Unlike a microwave oven, which concentrates its energy in pulses that cook food, the SPS extraordinary amount of energy is non-pulsated, and thus not dangerous to humans or the environment. As Sasaki puts it, "the beam wouldn't even be intense enough to heat your coffee" (ibid.). An experimental test has been proposed for 2018 to send a few kilowatts from low Earth orbit to the ground. JAXA then intends to demonstrate a prototype 100 kW SPS around 2020, followed by a 200 MW SPS. It would then seek to organize an international consortium to build a 1 GW commercial SPS beginning in the 2030s. Eventually, it is hoped, we will achieve O'Neill's goal of building a ring of SPS to provide Earth with much of the energy it will need in the centuries to come.

NASA is also paying attention to the possibility of abundant green energy from space, while advancing some novel initiatives of its own. During 2011–2012 the NASA Innovative Advanced Concepts (NIAC) Program supported the first phase of a project *SPS-ALPHA: The First Practical Solar Power Satellite* via *Arbitrarily Large Phased Array*. John C. Mankins, in the final report (2012) describes SPS-ALPHA as a collection of individually pointed thin-film mirrors in a geostationary Earth orbit (GEO) in a phased array with high-efficiency solid-state amplifiers in a very novel architecture (see picture below). The SPS would be built and maintained by autonomous robots. The development of this system would about 25 years, with a cost of about \$5 billion for a pilot plant and of \$20 billion for the first full-scale SPS (ibid, 106). Although this is lot of money (not even taking into account cost overruns), we should take into account that the International Space Station has cost over \$100 billion so far. A ring of SPSs based on this concept, however, "could deliver power on demand to more than 90 % of Earth's population at locations across the globe", with a near zero "carbon footprint" (ibid.). And it presumably could deliver all this at 10¢ per kW-h, or less, assuming improvements to some key technologies. This would be clean renewable energy at competitive prices (Fig. 3.5).

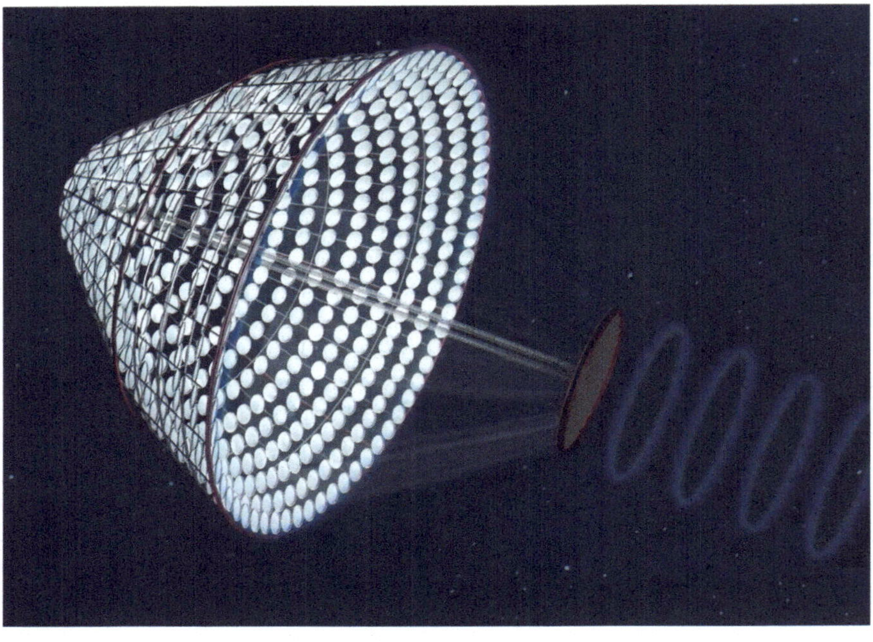

Fig. 3.5 One version of SPS-ALPHA. From Mankins (2012). Courtesy of NASA

I have offered these examples, in the voices of these new pioneers, so as to give the reader an impression of how vibrant the pursuit of last century's space dreams is becoming again. I could have offered many more. Apart from JAXA and NASA, the International Astronautical Federation (IAF) organizes the Space Power Symposium, annually. Thus we may expect many new and interesting proposals in the years to come.

It seems to me, however, that when we scale from one SPS to a ring of hundreds, let alone thousands of them in geosynchronous orbits, and perhaps many more in low-Earth orbits, we scale up in a way that is not likely to be benign to the environment. The excavation of that extraordinary amount of materials, the manufacturing of all those photovoltaic panels, cables, supporting structures, etc. are bound to create much pollution, and are not likely to offer a carbon footprint near zero. And then let's add the required mind-blogging number of rocket flights through the atmosphere to place all those materials in space. Such concerns bring us back to the wisdom of O'Neill's insistence in the mining of the necessary materials from the Moon and asteroids, and for the manufacturing of the SPSs in space. It appears, then, that we might be well advised to consider large space habitats where space builders will help solve some of our planet's critical problems while establishing a permanent human presence in space.

The British Interplanetary Society is responding to the call by launching its *Study Project Advancing Colony Engineering*, of which Jerry Stone is the project leader. In a somewhat parallel process to the initiatives to develop SPSs, the BIS's strategy also involves a pilot project, dubbed "Island Zero", on the way to realizing O'Neill's dream of an Island One. The aims of the Island Zero project include the following:

- Living under simulated gravity—something that has never been done before in space
- Construction from modular units and space-produced components
- Growing a large percentage of food in space
- Development of a space manufacturing facility. (BIS Communication.)

Their original proposal called for a connected ring of inflatable modules 5 m in diameter and 15 m long. Modules of that type have been launched by Bigelow Aerospace. The rather small radius (50 m), however, would require a high speed of rotation (4 rpm) to provide 1 g of artificial gravity. As of this writing, many of BIS members, and others, are suggesting a variety of structural designs, and exploring the physics and engineering of artificial gravity obtained through the rotation of the habitat, as well as the biology of closed biological environments so much smaller than the Earth.

3.5 Conclusion

We may then consider anew O'Neill's reasons for building artificial worlds. One important additional reason, implicit in the dispute I have discussed in this chapter, is that when considering our relationship with the environment we must turn to evolutionary biology. But evolutionary biology necessarily involves us with the matter of survival. And as O'Neill said, once a civilization achieves the capability of building space colonies, "it becomes unkillable in the physical sense" (1982, 190). The explanation is that we can move our space colonies around to colonize the solar system, and then perhaps even to expand into the galaxy. Space colonies may thus give our species the ability to escape the cosmic catastrophes (e.g., the expanding sun) that will eventually destroy life on Earth.[9]

References

Bernal, J. D. (1929). *The world, the flesh & the devil: An enquiry into the future of the three enemies of the rational soul.* Jonathan Cape.

Berry, W. (1977). In S. Brand (Ed.), *Space colonies.* Penguin Books.

Clark, W. (1977). In S. Brand (Ed.), *Space colonies.* Penguin Books.

Cockell, C. S. (2006). *Space on earth: Saving our world by seeking others.* MacMillan.

Ehricke, K. A. (1978). The extraterrestrial imperative. *Air University Review*, January–February. http://www.airpower.maxwell.af.mil/airchronicles/aureview_toc/AUReview1978/AUReview1978Jan-Feb.htm.

Glaser, P. (1968). Power from the sun: Its future. *Science, 162*(3856), 857–861.

Goto, D., Yoshida, H., Suzuki, H., Kisara, K., & Ohashi, K. (2014). *The overview of JAXA laser energy transmission R&D activities and the orbital experiments concept on ISS-JEM.* In Proc. international conference on space optical systems and applications (ICSOS) 2014, S5-2, Kobe, Japan, May 7–9, 2014. http://icsos2014.nict.go.jp/contents/pdf/S5-2.pdf.

Lewontin, R. C., Rose, S., & Kamin, L. J. (1984). *Not in our genes: Biology, ideology, and human nature.* New York: Pantheon Books.

Mankins, J. C. (2012). *SPS-ALPHA: The first practical solar power satellite via arbitrarily large phased array (A 2011–2012 NASA NIAC phase 1 project). Final report to NASA and NIAC.* http://www.nss.org/settlement/ssp/library/SPS_Alpha_2012_Mankins.pdf.

Meadows, D. (1977). In S. Brand (Ed.), *Space colonies.* Penguin Books.

Munévar, G. (2012). Self-reproducing automata and the impossibility of SETI. In A. Geppert (Ed.) *Imagining outer space*, Palgrave Macmillan, 267–284.

Noordung, H. (1929). *The problems of space flight.* Schmidt and Co.

Oberth, H. (1923). *Die Rakete zu den Planetenräumen* (By Rocket into planetary space). München, R. Oldenbourg.

O'Neill, G. (1982). *The high frontier.* New York: Anchor Press/Doubleday. First published in 1976.

[9]An earlier version of this chapter was presented at the conference "Envisioning Limits: Outer Space and the End of Utopia", in Berlin, April 19–21, 2012, organized by Alexander C.T. Geppert, Daniel Brandau and William R. Macauley. Much of the paper is excerpted from my book manuscript in progress, *The Dimming of Starlight: The Philosophy of Space Exploration.*

Sasaki, S. (2014). How Japan plans to build an orbital solar farm. http://spectrum.ieee.org/green-tech/solar/how-japan-plans-to-build-an-orbital-solar-farm.
Schwartz, J. S. J. (2011). Our moral obligation to support space exploration. *Journal of Environmental Ethics, 33*, 67–88.
Tsiolkowsky, K. (1903). *The rocket into cosmic space*. Naootchnoye Obozreniye: Moscow.

Chapter 4
Agonal Conflict and Space Exploration

Eleni Panagiotarakou

4.1 Introduction

In the film noir *The Third Man* (1949) Orson Welles, who plays the role of Harry Lime, is heard uttering the following line: "You know what the fellow said—in Italy, for thirty years under the Borgias, they had warfare, terror, murder and bloodshed, but they produced Michelangelo, Leonardo da Vinci and the Renaissance. In Switzerland, they had brotherly love, they had five hundred years of democracy and peace—and what did that produce? The cuckoo clock" (Ebert 2002:459). Granted that the above is historically inaccurate—the cantons were engaged in various civil and external conflicts, there was no democracy, and the cuckoo clock was a German invention—empirical evidence points towards a link between war, technological advancements, medical improvements and scientific progress (Donald 2014:151; Mosley 2011). In other words, Heraclitus' utterance "War is father of all and king of all"[1] is not entirely irrelevant in the case of war and space exploration.

War is said to be timeless, universal and inseparable from the human condition (Hanson 2010: xii). On the other hand, warfare (physical activities within the context of war) is in a continuous state of change as a result of military innovations. Military innovations are said to be driven by technology (Roland 2009), and technology is said to be driven by war (White 2005), leading to a plethora of innovations such as, computers, jet and airplanes (Restivo 2005). Apart from technological innovations, military funding has benefited numerous academic disciplines in the earth science such as oceanography, geodesy, seismology,

[1]*DK* 22B53 (Diels and Walther 1985).

E. Panagiotarakou (✉)
Department of Philosophy, Concordia University, Montreal, Canada
e-mail: eleni.panagiotarakou@concordia.ca

J.S.J. Schwartz and T. Milligan (eds.), *The Ethics of Space Exploration*,
Space and Society, DOI 10.1007/978-3-319-39827-3_4

astrophysics, geography and biotechnology, to say nothing of the social science such as, psychology and political science.[2] However, no field has benefitted more greatly from war than space exploration. The space race (ca 1955–1972) was made possible by unprecedented public funding by the United States and the Soviet Union in the context of the Cold War.

The intellectual resources that enabled spaceflight came in the form of rocket scientists from the Third Reich. As WWII came to an end there were urgent efforts by Britain, US and the Soviet Union to capture the rocket scientists of Nazi Germany. This included scientists who were working in the design of the V-2 rocket, namely, Walter Dornberger, Konrad Dannenberg and Wernher von Braun.[3] In the specific case of America, a secret program, "Exploitation of German Specialists in Science and Technology in the United States" (later to be renamed Operation Paperclip), saw the relocation of hundreds of German rocket and aeronautic scientists with the objective of enhancing that country's technological and scientific capabilities. The Soviets run a similar program by the name of Operation Osoaviakhim which likewise extracted hundreds of rocket and aeronautic scientists from Germany.[4] In other words, spaceflight arose from one of humanity's largest conflict, WWII.

In the early days space funding was meagre. For example in 1947 only $47 million was allocated to missile development in the United States (Brzezinski 2007:87).[5] However, following deteriorating political relations between the USSR and the USA in the summer of 1946, funding increased to substantial levels. In the specific case of the USA, and following a lethargic start with the Vanguard Project 1955–1956 (which suffered as a result of inadequate funding), the impetus was the Soviet's successful launch of the *Sputnik 1* into the Earth's orbit (Garber and Launius 2005). With the Space Race underway one of the first actions by the American government was the establishment of the National Aeronautics and Space Administration (NASA) by President Dwight D. Eisenhower in 1958.

---

[2]One of the consequences was the development of a close relationship between defence departments and academia reflected in the term "military-industrial-academic complex" (Kevles 1971), as opposed to simply the "military-industrial complex" (Koistinen 1987; Pursell 1972).

[3]For a balanced view of von Braun see Neufeld (2007). Unlike Piszkiewicz (1998) who portrays Von Braun as a Nazi villain who escaped the wheels of justice, Neufeld suggests that Von Braun was driven by his passion for space flight which, in turn, led him to a "Faustian bargain" with the Third Reich in order to get access to needed resources. If true this would put Von Braun in the same category as natural philosophers such as, Thales of Miletus (624 BC–546 BC) and Archimedes (287 BC–212 BC) who were motivated by their intellectual curiosities. For example, Archimedes was said to be so taken with a geometrical problem that he was oblivious to the fall of his town, Syracuse. When a soldier was sent to fetch him on the orders of the Roman General Marcus Claudius Marcellus he dismissed him with his famous line "Noli turbare circulos meos!" (Do not disturb my circles!).

[4]For an in-depth discussion of this program see Maddrell (2005: 173–206) and Naimark (1995).

[5]Consider Wernher Von Braun's half-joking claim that [at] "Peenemünde [German municipality where the V-2 rocket was developed] we'd been coddled. Here [US] they were counting pennies" (Brzezinski 2007: 87).

NASA began with 8000 employees, a $100 million annual budget, and numerous research laboratories. Apart from its initial absorbing of the National Advisory Committee for Aeronautics, NASA went on to incorporate other organizations such as the Jet Propulsion in California and the Army Ballistic Missile Agency in Alabama (ibid). While many of NASA's initial projects were civilian (e.g., information-gathering satellites for environmental monitoring, meteorology) others were not. Programs geared towards space exploration included the Mercury 1959–1963 (feasibility/functionality of humans in spacecraft); the Gemini 1961–1966 (development of spaceflight capabilities); the Apollo 1969–1972 (Moon exploration); and Skylab 1973–1979 (space station) projects. Later day projects included the Space Shuttle (spaceflight/space transportation), the Pioneer and Voyager (interplanetary space travel), and the Mars Pathfinder (planetary exploration). Comparable Soviet programs included, Voskhod 1965 (spaceflight); Luna 1959, 1966, 1970 (lunar exploration); and Venera 1961, 1970 (planetary exploration) among others (Heppenheimer 1997).

While it is beyond the scope of this chapter to delve into a detailed discussion of the above mentioned programs, it will suffice to say that in a relatively very short time an astonishing amount of scientific and technical progress was made in the areas of spaceflight, lunar exploration, and planetary exploration. Put differently, an unprecedented accomplishment in the history of humankind.

At the risk of repetition, what made this feat possible was the unique geopolitical environment—a bipolar distribution of power in the international arena. The USA and the USSR utilized the spaceflight and planetary exploration programs as an assertion of superiority. What made this conflict extraordinary was the fact that it was a non-violent war. To put it in the words of Margaret Thatcher, it was a war which was won "without firing a shot."[6] While some would say that Thatcher's statement is an hyperbole—especially when we take into consideration the casualties of numerous proxy wars (i.e., Korea, Vietnam, Afghanistan and Cambodia to mention but a few)— the fact remains that the Cold War entailed few casualties for the principal actors.[7] While there was a dangerous buildup of biological and nuclear weapons the doctrine of Mutually Assured Destruction (MAD) managed to prevail despite or because some close calls such as the Cuban Missile Crisis.[8]

Insofar as the Cold War was nonviolent, and insofar as it prompted the USA and USSR to engage in the Space Race, it would seem to have resembled the ancient Greek spirit of agon whereby the objective was not to annihilate one's opponent but to surpass them in a struggle for excellence. The concept of agon is first articulated in Hesiod's work where it is expressed via the narrative genre of mythopoesis. Hesiod distinguishes between two goddesses bearing the same name: Eris. The two

---

[6]The entire quote reads: "He won the Cold War without firing a shot, but it was because he pointed like a thousand missiles at Soviet cities" and was made in reference to the American President Ronald Reagan in 1991 at a Heritage Foundation dinner (D'Souza 1997:23).

[7]With thanks to Janice Freamo and Stefan Dolgert for their feedback on this point.

[8]For a discussion of four other close encounters see Forden (2001).

goddesses, Hesiod tells us, possessed two distinct natures. That distinction is worth quoting here at length.

> So, after all, there was not one kind of Strife alone, but all over the earth there are two. As for the one, a man would praise her when he came to understand her; but the other is blameworthy: and they are wholly different in nature. For one fosters evil war and battle, being cruel: her no man loves; but perforce, through the will of the deathless gods, men pay harsh Strife her honour due. But the other is the elder daughter of dark Night and the son of Cronos who sits above and dwells in the aether, set her in the roots of the earth: and she is far kinder to men. She stirs up even the shiftless to toil; for a man grows eager to work when he considers his neighbor, a rich man who hastens to plough and plant and put his house in good order; and neighbour vies with his neighbour as he hurries after wealthy. This Strife is wholesome for men. And potter is angry with potter, and craftsman with craftsman, and beggar is jealous of beggar, and minstrel of minstrel (*Works and Days*, 11–24).[9]

In *Homer's Contest* Nietzsche provides an equally powerful analysis of the same concept albeit from an artistic, literary and dramatic context. Closer to our time, the competitive but non-destructive Eris that Hesiod and Nietzsche alike describe, personified the essence of the Space Race.

4.2 The End of the Cold War and the Decline of NASA's Funding

This is not to say that space scientists—Soviet and American alike—were engaged in a pure and unadulterated scientific quest for cosmic knowledge while blissfully unaware of the surrounding political and military realities. As already mentioned, the majority of rocket scientists came from the Third Reich and the Space Race was the offspring of the Cold War. In addition, recently declassified documents confirm one of the worst-kept secrets: NASA was collaborating with the Central Intelligence Agency (CIA) and the Department of Defence (DoD) (David 2015). The same documents also reveal that NASA's cooperation with the CIA and the DoD was often strained,[10] suggesting that NASA scientists engaged in such collaborations from a sense of Thucydidean *ananke* (necessity) brought about by financial constraints.

The end of the Cold War resulted in draconian cuts to space programs on both sides of the Atlantic. In the specific case of the Russian space program—the

---

[9]As an interesting side note it should be noted that a newly discovered dwarf planet in the Kuiper Belt was named 'Eris' (after the goddess of strife and war). Fittingly, Eris' moon was named 'Dysnomia' (lawlessness) by the astronomer who discovered both of them (Tytell 2006).

[10]Mostly over cost sharing but other areas of contention included "access to classified information, encryption of data originally intended for civilian use, and delays to military satellite launches caused by the *Challenger* disaster" (The National Security Archive 2015).

successor of the Soviet space program—news reports suggest a full-blown crisis (Chance 2015). Apart from corruption scandals (a plague in modern-day Russia) however, the main culprit appears to be the same as in the case of NASA: a diminished sense of urgent national needs in a post-Cold War environment (Logsdon 2007). With that threat gone the justification for the continuing existence of space programs stems mainly from the following three factors: (1) environmental/civic programs (radar topography, oceanographic satellites), (2) commercial endeavours (communications satellites), and (3) space exploration for the love of knowledge, that is to say, without any tangible benefits.

In the case of NASA, none of the aforementioned factors have stopped the decline of state funding.[11] NASA officials have been relentlessly lobbying the Congress for increased funding for many years. The latest, and rather passionate, manifestation took place in April 2014 when NASA Administrator Charlie Bolden was quoted as saying that it was his intention to "get down on my hands and knees and beg and plead and make [Congress] understand this country needs human access to space" (Clark 2014).

Efforts to raise public awareness have been met with little success. A rather tragicomic example is the recent car wash and cupcake giveaway by planetary scientists at NASA's Jet Propulsion Laboratory in Pasadena, California. Designed to raise awareness to the proposed 2013 budget cuts (21 % from the planetary budget and 38 % from the Mars projects), people were given free cupcakes in exchange for signatures to a petition "beseeching Congress to reverse the [governmental] cuts" (Dance 2012). Judging from the latest round of austerity measures —$250 million from NASA's Earth science program on May 2015 (Foust 2015)— it would seem that the free cupcakes failed to curb the appetite of Congress for funding cuts. A similar scenario is unfolding in Russia and Canada which are also facing considerable governmental cuts (CBC News 2012).

The reason for the failure to win popular support lies (mostly) with perceived priorities.[12] Simply put, money for space exploration is seen as wasted money. Popular opinion holds that, given the number and gravity of issues here on earth such as, poverty, education, global population growth, regional wars, species extinction, and environmental degradation, tax money would be better spent

[11]Mick (2014) data derived from the Augustine Report, NASA History, http://history.nasa.gov/augustine/racfup5.htm Accessed June 5, 2015. Also useful the Budgetary Implications of NASA's Current Plans for Space Exploration (Congressional Budget Office 2009) http://www.cbo.gov/sites/default/files/04-15-nasa.pdf. According to John Hickman the fact that "NASA's budget clearly collapsed with the end of the First Cold War under JFK and LBJ is indisputable…and can be explained as Nixon's revenge on JFK and the 1967 Outer Space Treaty's elimination of the most important payoff for states that engage in exploration: new sovereign territory. That NASA's budget did not recover with the Second Cold War under Reagan doesn't mean we weren't spending on space but instead that a lot of the funding was directed through the DoD rather than NASA" (electronic communication, January 5, 2016).

[12]Steinberg (2011) points out in an insightful article that public opinion is united on the topic of space exploration (desirable) but divided on its funding (too high). The government's response? Each year give NASA more money but at the same time a smaller percentage of the federal budget.

addressing those issues. If money is to be given to NASA, one popular argument holds, they [NASA] should use that funding for earth-based space projects. Or, as vocal critic Gerard J. DeGroot puts it, "Obscenely expensive manned missions mean that practical, earth-based science suffers, as does the genuinely valuable satellite research so essential to the way we live today" (2009).[13]

It should be noted that reluctance to fund space programs is not a recent phenomenon. The large costs of the space program made it a target even prior to the end of the Cold War. In the case of the USA one of its earliest opponents was President Dwight Eisenhower—despite or because of the fact that he was the founder of NASA. Fearing uncontrollable financial spending with no end in sight Eisenhower argued that "every rocket fired signifies, in the final sense, a theft from those who hunger and are not fed, those who are cold and are not clothed."[14] Granted that Eisenhower was also referencing the gargantuan costs of the armaments race, nonetheless, variations of his argument survive to this day (more on this shortly).

4.3 Exploring Alternative Funding Sources

In response to reduced budgets a number of alternatives have been explored. In the specific case of NASA this includes the following non-exhaustive list: cooperation and partnerships with other space exploration entities with the aim of pooling together financial and technical resources (Augustine et al. 2009: 7); increasing efficiency by eliminating wasteful and duplicative programs (Chapman 2015); expanding outreach programs beyond the traditional support base of young, white, educated, well-to-do Republicans (Cobb 2011); space tourism private space exploration (Giancarlo 2014; Solomon 2008); and managing NASA on a "market-based, competitive, performance-oriented principles" (Heracleous and Gonzalez 2014).

In addition to the above there is a growing chorus of voices calling for lunar, asteroid and planetary mining. Initially the topic of science fiction novels (recall Isaac Asimov's *Catch That Rabbit*), private companies the likes of Planetary Resources, Deep Space Industries (DSI), Virgin Galactic, Blue Origin, and Space Exploration Technologies Corporation (SpaceX) are now courting NASA. While some argue that this is pure speculation—theoretically possible but economically unviable (Crawford 2015)—asteroid mining will, in all likelihood, take place in the coming decades as a result of improved infrastructure (Sanchez and McInness 2012). Aiding this process will be legal guarantees allowing private companies ownership of asteroid resources thereby ensuring profitability, to say nothing of attracting greater levels of venture capital. That certainly seems to be the goal of the "American Space Technology for Exploring Resource Opportunities in Deep Space Act" or the "ASTEROIDS Act" (U.S. Congress. House. Committee on Science,

[13]For a somewhat similar critique see Etzioni (2012).

[14]The above was part of a speech titled *The Chance for Peace* that was delivered to the American Society of Newspaper Editors in Washington on April 16th 1953 (Bury 2013: 47–48).

Space, and Technology 2014). Indeed, in a rare display of bipartisanship, Republicans and Democrats recently passed the "Commercial Space Launch Competitiveness Act" (CSLCA) legislation which was signed in November 2015 by US President, Barack Obama.[15]

Leaving aside the controversial political aspects of this bill—e.g., CSLCA basically gives private American companies free rein in space by allowing them to (a) bypass regulatory oversight until 2023, and (b) own the resources they obtain—there are those who argue that commercial entities cannot bear the full cost of extractive activities in space. For instance, Mann suggests that private American companies need NASA's cooperation to be financially viable despite having the financial backing of entrepreneurial billionaires the likes of Larry Page (CEO, Google) and Charles Simonyi (Chief Architect, Microsoft), to mention but a few. Indeed, Neil deGrasse Tyson's recent comments that private companies won't be able to take the lead in space exploration must be understood in this context; after all the first Europeans to come to America were not the Dutch East India Company but Christopher Columbus whose expedition was sponsored by Spain (Ha 2014). The above is best understood within the context of what Hickman (1999) calls the "problem of capitalization". Large amounts of capital, a high risk environment, and long-term waiting periods for economic profits amounts to nothing less than an insurmountable obstacle. According to the same author, governments are not better at managing gargantuan projects than private firms [according to Taleb (2012) they are actually worse] but they are better at financing them.[16]

Questions of legality and private sector economic viability aside, however, there are environmental concerns associated with space mining (Merchant 2013). In the specific case of lunar mining, there is the simultaneous depletion of fossil fuels and the rising of CO_2 levels on earth—to say nothing of the moon's cultural significance to humanity (Milligan 2015). More importantly, however, "near-Earth resources" (namely, resources located on the moon and nearby asteroids) are scarce and non-renewable (Schwartz and Milligan 2016) . Helium-3 concentrations in the lunar regolith are shallow and geographically differentiated, while asteroid regoliths are likely to contain lower concentrations of the same gas (ibid). Notwithstanding futuristic technologies that would be capable of interstellar travel and efficient harvesting techniques—and here visions of the "heighliners" and "harvesters" from Frank Herbert's science fiction novel *Dune* spring to mind[17]—scarce "near-Earth resources" might be humanity's sole reality for decades if not centuries.

---

[15]U.S. Congress. House. Committee on Science, Space, and Technology, "A. 1297—U.S. Commercial Space Launch Competitiveness Act." https://www.congress.gov/bill/114th-congress/senate-bill/1297.

[16]For a detailed discussion, including a sophisticated theoretical explanation, in addition to Hickman (1999) see also Hickman and Dolman (2002).

[17]For those not familiar with this popular science fiction work, the "harverster" was a large, mobile machine that was used to harvest a rare and valuable resource by the name "Spice Melange" from the surface of a desert planet. This substance enabled the Guild Navigators (mutated humans) to

The non-renewable nature of Helium-3 in the lunar regolith, however, has not stopped the chief scientist of the China Lunar Exploration Project (CLEP), Ouyang Ziyuan, from stating on multiple occasions that the principal goal of the Chinese space program would be lunar mining for Helium-3 (Reynolds 2010). Indeed, China's growing obsession with a space program can be seen by its annual launch rate which has now surpassed those of Europe and the United States (Harvey 2013). Indeed, it is predicted that by 2020 China will have their own space station and will soon rise to be the "dominant player in human space exploration" (Berger 2013). China is not only matching Russia and the USA but, far more importantly, it's reaching the same milestones in less time and while avoiding costly, financial mistakes (Hickman 2013).

Some see the rise of China as a manifestation of the rise of multipolarity in the international space landscape (Verschuuren 2011). However, if China were to continue her space program with the same rising trajectory, and while accepting the premise that the current international system is one of tripolarity involving USA/China/Russia (Hall 2014), the likelihood of a bipolar space order is quite high. Leaving the European Space Agency aside (on account of its diminishing funding as a result of the euro crisis (Messier 2011; Randall 2011) if Russia's space program suffers any further technological and financial setbacks, China and the USA will emerge as the undisputed hegemons of space. At that stage the question to ponder would be whether or not the USA government would increase NASA's funding to avoid falling behind China. In this regard the recently passed "Commercial Space Launch Competitiveness Act" (CSLCA) legislation could be viewed as the first step towards protecting private American space mining interests against Chinese state interests. Whether this will result in geopolitical tensions is uncertain. For example, while Obama signed the CSLCA bill he also expressed an interest in cooperating with China (Beldavs 2015). Should the future relationship between the USA and China become strained, will that lead to an agonistic environment in relation to space policy? Again, there is much uncertainty.

What is certain however is that Chinese state interests and private American corporations are driven by a quest for raw materials and their intended purpose is to meet future energy and consumption needs. Some aspects of this purpose, such as the greening of the earth (Matloff et al. 2014), are noble. Others, such as the replenishment of raw materials that are being quickly exhausted due to unsustainable consumption patterns and/or uncontrollable human population growth, are not. Their lack of nobility stems from their enabling status. That is to say, lunar and asteroid raw materials are likely to enable the continuation of unsustainable consumption patterns and overpopulation patterns. As a matter of fact, if past technological innovations are any indication [i.e., the agricultural revolution, the industrial revolution, and the green

(Footnote 17 continued)

safely navigate through interstellar space the "heighliners"—immense starships used for the transportation of people and equipment (Herbert 1965).

revolution (Rogers et al. 2012)], space mining might lead to a new surge in population growth. Last, but not least, the main (and often sole) motive is the generation of monetary profits. However, as I will argue in the following section, wealth-generation, philosophy and astronomy are antithetical.

4.4 The Heavens, Money and the Philosophical Gaze

In Plato's *Phaedo* (96a6-8) proto-philosophy is described as "that wisdom which people call the inquiry regarding nature". In other words, natural philosophy—the precursor of modern natural science. According to Aristotle the founder of natural philosophy was Thales of Miletus (*Metaphysics*, 983 b21–22).[18] Unlike Babylonians priests Thales did not turn his gaze towards the celestial bodies for the purpose of astrology—a pseudoscience purporting a link between astronomical phenomena and human affairs (Pigliucci 2010:63). Thales looked to the heavens for epistemological knowledge. We know of two apocryphal but interesting ancient tales about Thales of Miletus. The first tale tells of a mishap; falling into a pit. The second tale tells of an entrepreneurship; making huge profits during a monopolized business model. Both stories are relevant to the topic of space exploration for reasons that will soon become apparent.

In the first tale, Thales is depicted in Plato's *Theaetetus* (174a) as a star-gazing philosopher who falls into a pit and becomes the object of laughter by a "witty Thracian servant girl." Prior to Plato's *Theaetetus* the image of an astronomer falling into a well is first encountered as an Aesopian fable. However, Aesop's fable makes no mention of the laughter of the passer-by, their sex (Glasgow 2009:12), their social status, or their ethnicity.[19] That Plato would modify the fable and associate laughter at the expense of a natural philosopher in the figure of a non-Greek, slave, woman, that is to say, the lowest denominators in 4th century Athens, is also not without significance.[20] That both Heidegger and Arendt would quote the same passage but arrive at differing conclusions—Heidegger seeing only

---

[18]The term 'philosophia' (as opposed to individual 'philosophos') did not appear until after Thales' time, and 'philosophy' per se was not unique to Greece—think of the Upanisads in India. With thanks to Gary Shapiro on the evolution of the term and Patrick Lee Miller on its non-exclusiveness (electronic communication, December 22, 2014). For an in-depth analysis of both points see Geoffrey Lloyd, "Philosophy: What Did the Greeks Invent and is it Relevant to China?" *Extrême-Orient, Extrême-Occident* 27 (2005): 149–159 and Patricia Curd, "Presocratic Philosophy," *The Stanford Encyclopedia of Philosophy* (Winter 2012), ed. Edward N. Zalta http://plato.stanford.edu/archives/win2012/entries/presocratics.

[19]While a number of fables are attributed to Aesop this does not imply that these fables belong to the historical figure of Aesop (if indeed he existed). Aesop has as much to do with "fables ascribed to him as King David with the Psalms, King Solomon with the proverbs, and Joe Miller with the jests" (Leofranc Holford-Strevens, electronic communication).

[20]According to Zafiropoulos (2001) Plato did not perceive Aesopic fables as being inferior as seen by his willingness to utilized them in order to elucidate his arguments.

"the laughter of the uncomprehending non-philosopher,"[21] in contrast to Arendt's more subtle analysis of an intramural warfare between thought and common sense which philosophers have interpreted "as the natural hostility of the many and their opinions towards the few and their truth" (Arendt 1978, 81–82)—is also not without significance. This significance is best understood within the context of the second tale which is recounted by Aristotle and is worth quoting in its entirety.

> The story goes that when they [fellow citizens] were reproaching him for his poverty, supposing that philosophy is useless, he learned from his astronomy that the olive crop would be large. Then, while it was still winter, he obtained a little money and made deposits on all the olive presses both in Miletus and in Chios, and since no one bid against him, he rented them cheaply. When the time came, suddenly many requested the presses all at once, and he rented them out on whatever terms he wished, and so he made a great deal of money. In this way he proved that philosophers can easily be wealthy if they wish, but this is not what they are interested in (*Politics* 1.11 1259a9-18).

Placing the above in the contemporary context, it becomes obvious that ancient prejudices die hard. Space science is praised for practical outcomes such as, satellite communications and space-based navigations systems but rarely for the discoveries of exoplanets. The fact that NASA has published a staggering 1,800 reports on positive spinoffs since 1975 (NASA Spinoff 2015) is a testament to political pressures originating in the electorate. In essence, warfare eliminates the problem of justification.[22]

Similar to the ancient Miletians who were pressuring Thales to demonstrate the practically of his science, American administrations—Republican and Democrat alike—have been pressuring NASA to justify its existence in terms of tangible, practical benefits. Failure to demonstrate "practical benefits" risks adding to the "myth of wasted taxpayer dollars" (ibid). NASA, the argument goes, must prove how it ultimately benefits "the American Consumer" (ibid). It would seem to me that, those who support the funding of space science solely on economic and societal benefits fall victim to a variant of the same "practical mentality" as the ancient Miletians. Under this mentality only one eye is allowed to gaze into the heavens, the other must keep its gaze on the earth.

---

[21]In all fairness to Heidegger his claim "Philosophy, then, is that thinking with which one can start nothing and about which housemaids necessarily laugh. Such a definition of philosophy is not a mere joke but is something to think over. We shall do well to remember occasionally that by our strolling we can fall into a well whereby we may not reach ground for quite some time" (Taminiaux 1997, 2).

[22]With thanks to Janice Freamo on this point.

4.5 Conclusion

In this chapter I have argued that one of the unintended, positive consequences of war is significant technological advancements and scientific progress. In the specific case of the Cold War the same benefits took place without the negative consequences of violent conflict—death and suffering. I have also argued that the unique environment of the Cold War resembled the ancient agon whereby the objective was to surpass one's opponent in excellence without destroying them. The lucky recipient of that historical agonistic moment was spaceflight and space exploration.

The end of the Cold War saw draconian funding cuts to international space programs including NASA's. Sadly, although not surprising, space programs are being placed at the bottom of budgetary priorities because they are seen both as non-urgent and non-pragmatic. In the case of NASA, the most promising source of new funds are private companies in search of financial profits. However, there are two potential problematic aspects to such a partnership. The first aspect is insufficient funding—a feasible space program needs the unlimited financial resources of a wealthy nation-state. This means that despite the recent favourable legislation in the form of the "Commercial Space Launch Competitiveness Act" (2015), long-term viability might prove elusive.

The second problematic aspect is a contradictory value-systems of NASA and the various private companies. Assuming that the values of NASA scientists reflect those of their ancient founder, Thales of Miletus, they would include such values as scientific knowledge for its own sake, expanding the intellectual horizons of humankind, and gaining a greater appreciation of the universe. These values are antithetical to those of private companies which are mostly headed by venture capitalists whose sole objective is the maximization of profits. Put differently, and taking a page from Plato's *Republic*, the former are lovers of wisdom while the latter are lovers of money.

References

Arendt, H. (1978). *The life of the mind*. Orlando: Harcourt Inc.

Aristotle. (1932). *The Politics*, with an English translation by H. Rackham. London: W. Heinemann.

Augustine, R. Norman et al. (2009). *Seeking a human spaceflight program worthy of a great nation*. Executive summary. Review of U.S. Human spaceflight plans. Washington, D.C., NASA: Committee Office of Science and Technology Policy.

Beldavs, V. (2015). Prospects for US-China space cooperation *The Space Review*. December 7. http://www.thespacereview.com/article/2878/1

Berger, E. (2013). As China's space program rises, will NASA rise again? *Houston Chronicle*, November 05, http://www.houstonchronicle.com/news/nation-world/article/As-China-s-space-program-rises-will-NASA-rise-4958549.php

Brzezinski, M. (2007). *Red moon rising: Sputnik and the hidden rivalries that ignited the space age*. New York: Henry Holt.

Bury, H. (2013). *Eisenhower and the military-industrial complex: 'Open Skies' and the Cold War arms race*. London: I.B. Tauris.

CBC News. (2012). Canadian space programs face budget cuts, layoffs. April 09, http://www.cbc.ca/news/technology/canadian-space-programs-face-budget-cuts-layoffs-1.1182669

Chance, M. (2015). Russia's space program in crisis? *CNN News*, June 3, http://www.cnn.com/2015/06/03/europe/russia-space-crisis-chance/

Chapman, B. (2015). Waste and duplication in NASA programs: The need to enhance U.S. space program efficiency. *Space Policy, 31*, 13–20.

Clark, S. (2014). NASA would get funding hike under congressional budgets. *Spaceflight Now*. June 5, http://spaceflightnow.com/news/n1406/05appropriation/#.VXJEoSzbJjo

Cobb, W. W. (2011). Who's supporting space activities? An 'issue public' for US space policy. *Space Policy, 27*, 234–239.

Crawford, I. (2015). Lunar resources: A review. *Progress in Physical Geography, 39*, 137–167.

Dance, A. (2012). NASA scientists fight budget cuts with cupcakes: Planetary researchers bake cakes and shine shoes to raise awareness of declining budget. *Nature*, June 11. http://www.nature.com/news/nasa-scientists-fight-budget-cuts-with-cupcakes-1.10805

David, E. J. (2015). *Spies and Shuttles: NASA's Secret Relationships with the DoD and CIA*. Gainesville: University Press of Florida.

DeGroot, J. (2009). The space race is a pointless waste of money. *The Telegraph*, February 25, http://www.telegraph.co.uk/comment/personal-view/4807506/The-space-race-is-a-pointless-waste-of-money.html#disqus_thread

Diels, H., & Walther, K. (1985). *Die Fragmente der Vorsokratiker. 2 vols* (6th ed.). Zurich: Weidmann.

Donald, G. (2014). *Lies, damned lies and history: A catalogue of historical errors and misunderstandings*. Stroud: The History Press.

D'Souza, D. (1997). *Ronald Reagan: How an ordinary man became an extraordinary leader*. New York: Touchtone.

Ebert, R. (2002). *The great movies* (Vol. 1). New York: Broadway Books.

Etzioni, A. (2012). Mars can wait. Oceans Can't. *CNN News*, August 17, http://www.cnn.com/2012/04/09/opinion/etzioni-space-oceans/

Forden, G. (2001). *Reducing a common danger: Improving Russia's early-warning system* (p. 399). Cato: Policy Analysis, no.

Foust, J. (2015). House budget cuts NASA earth science by more than $250 Million. *Space News*, May 19, http://spacenews.com/house-budget-cuts-nasa-earth-science-by-more-than-250-million/

Garber, S. J., & Launius, R. D. (2002). A brief history of the national aeronautics and space administration. In C. V. Anderson (Ed.), *National aeronautics and space administration*. New York: Nova Science Publishers Inc.

Giancarlo, G. (2014). Private space exploration: A new way for starting a space-faring society? *Acta Astronautica, 104*, 480–486.

Giroux, A. H. (2007). *The University in Chains: Confronting the military-industrial-academic complex*. Boulder: Paradigm Publishers.

Glasgow, R. (2009). *The concept of water*. Glasgow Books.

Ha, A. (2014). Neil deGrasse Tyson says private companies won't take the lead in space exploration. *TechCrunch*, March 8, http://techcrunch.com/2014/03/08/neil-degrasse-tyson-sxsw/

Hall, O. G. (2014). *Authority, ascendancy, and supremacy: China, Russia, and the United States*. New York: Routledge.

Hanson, D. H. (2010). *The father of us all: War and history, ancient and modern*. New York: Bloomsbury.

Harvey, B. (2013). *China in space: The great leap forward*. New York: Springer.

Heppenheimer, T. A. (1997). *Countdown: A history of space flight*. New York: Wiley.

Heracleous, L., & Gonzalez, S. (2014). Two modest proposals for propelling NASA forward. *Space Policy, 30*, 190–192. doi:10.1016/j.spacepol.2014.08.008.

Herbert, F. (1965). *Dune*. Radnor: Chilton book company.
Hesiod. (2008). *The Homeric Hymns, and Homerica: Hesiod and Homer*. Translated by Hugh G. Evelyn-White. Stilwell: Digireads.com Publishing.
Hickman, J. (1999). The political economy of very large space projects. *Journal of Evolution and Technology, 4*, 1–14.
Hickman, J. (2013). China is winning the space race. *Foreign Policy*. August 3, http://foreignpolicy.com/2013/08/03/china-is-winning-the-space-race/
Hickman, J., & Dolman, E. (2002). Resurrecting the space age: A state-centered commentary on the outer space regime. *Comparative Strategy, 21*, 1–20.
Koistinen, C. P. (1987). *The military industrial complex: A historical perspective*. New York: Random House.
Logsdon, M. J. (2007). Space in the post-cold war environment. In S. J. Dick & R. D. Launius (eds), *Societal impact of spaceflight* (pp 90–102). Washington: NASA History Division, Office of External Relation.
Maddrell, P. (2005). Operation matchbox and the scientific containment of the USSR. In P. Jackson & J. Siegel (Eds.), *Intelligence and statecraft: The use and limits of intelligence in international society*. Westport: Praeger.
Matloff, G., Bangs, C., & Johnson, L. (2014). *Harvesting space for a greener earth*. New York: Springer.
Merchant, B. (2013). The problem with asteroid mining. *Vice News* January 23, http://motherboard.vice.com/blog/the-problem-with-asteroid-mining
Messier, D. (2011). Euro debt crisis roils space efforts. *Parabolic Arc*, November 10, http://www.parabolicarc.com/2011/11/10/euro-debt-crisis-roils-space-efforts/
Mick, J. (2014). U.S. National Research Council NASA's plan to reach Mars will fail. *Daily Tech News*. June 11, http://www.dailytech.com/US+National+Research+Council+NASAs+Plan+to+Reach+Mars+Will+Fail+/article35013.htm. Accessed June 5, 2015.
Milligan, T. (2015). *Nobody owns the moon: The ethics of space exploitation*. Jefferson: McFarland & Company Inc, Publishers.
Mosley, M. (2011). How war has driven medical advances. BBC News Website. http://www.bbc.com/news/health-15771688
Naimark, M. N. (1995). *The Russians in Germany: A history of the Soviet zone of occupation, 1945–1949*. Cambridge: Belknap Press.
NASA. History of NASA. Augustine report. http://history.nasa.gov/augustine/racfup5.htm
NASA. *Spinoff*. https://spinoff.nasa.gov/index.html
Neufeld, J. M. (2007). *Von Braun: Dreamer of space, engineer of war*. New York: A.A. Knopf.
Pigliucci, M. (2010). *Nonsense on stilts: How to tell science from bunk*. Chicago: University of Chicago Press.
Piszkiewicz, D. (1998). *Wernher Von Braun: The man who sold the moon*. Westport: Praeger.
Pursell, W. C. (1972). *The military-industrial complex*. New York: Harper & Row.
Randall, C. (2011). Crisis could end space-age dreams for European Space. *The National*, October 5, http://www.thenational.ae/business/industry-insights/technology/crisis-could-end-space-age-dreams-for-european-space-agency
Restivo, S. (2005). *Science, technology, and society: An encyclopedia*. Oxford: Oxford University Press.
Reynolds, N. (2010). Many moons to go: The promise of lunar mining. *The Globe and Mail*, July 5, http://www.theglobeandmail.com/globe-debate/many-moons-to-go-the-promise-of-lunar-mining/article1386467/
Rogers, P. P., Kazi F. J., & J. A. Boyd. (2012). *An introduction to sustainable development*. UK: Earthscan Publications Ltd.
Roland, A. (2009). "War and technology" Foreign policy research institute. http://www.fpri.org/articles/2009/02/war-and-technology
Sanchez Cuartielles, J.-P., & McInness, C. (2012). Assessment of the feasibility of future shepherding of asteroid resources. *Acta Astronautica, 73*, 49–66.

Schwartz, S. J. J., & Tony, M. (2016). Some ethical constraints on near-earth resource exploitation. *Yearbook on Space Policy 2015*. New York: Springer (forthcoming).

Solomon, L. (2008). *The privatization of space exploration: Business, technology, law and policy*. New Brunswick, NJ: Translation Publishers.

Steinberg, A. (2011). Space policy responsiveness: The relationship between public opinion and NASA funding. *Space Policy, 27*, 240–246.

Taleb, N. N. (2012). *Antifragile: Things that gain from disorder*. New York: Random House.

Taminiaux, J. (1997). *The Thracian maid and the professional thinker: Arendt and Heidegger*. Edited by Michael Gendre. Albany: SUNY Press.

The National Security Archive. (2015). NASA's secret relationships with U.S. defense and intelligence agencies. http://nsarchive.gwu.edu/NSAEBB/NSAEBB509/. Accessed April 13, 2015.

Tytell, D. (2006). All hail Eris and Dysmonia. *Sky & Telescope*, September 14. http://www.skyandtelescope.com/astronomy-news/all-hail-eris-and-dysnomia/

U.S. Congress. House. Committee on Science, Space, and Technology. Subcommittee on Space (2014). *American Space Technology for Exploring Resource Opportunities in Deep Space Act*. 113th Cong., H.R. 5063.

Verschuuren, P. (2011). Geopolitique spatiale: vers une course a l'espace multipolaire? *Revue intenationale et strategique, 4*(84), 40–49.

White, M. (2005). *The fruits of war: How military conflict accelerates technology*. New York: Simon & Schuster Ltd.

Zafiropoulos, C. (2001). *Ethics in Aesop's fables: The Augustana collection*. Leiden: Brill.

Chapter 5
Prospects for Utopia in Space

Christopher C. Yorke

5.1 Models of Utopia

I have tasked myself with analyzing the prospects for a utopia being achieved in space. In order to accomplish this, we must first look at the wide variety of ways that utopia has been conceptualized on this planet, in the numerous academic fields in which it finds application. Making no claims of exhaustiveness, I will consider three importantly distinct models of the concept of utopia: (1) the teleological model, (2) the discursive model, and (3) the horizonal model.

A teleological model of utopia is any vision which posits an ideal end-state for a society; a final resting-place wherein perfection or, failing that, a state of socio-political optimality is achieved for its citizenry. Some utopians may claim that the purpose or culmination of human history is the achievement of utopia, or that utopia is contingently or logically inevitable, although such strong premises are not necessary to ground the teleological model. Its sphere of application is chiefly political, in motivating action to achieve a desired picture of the world. Teleological utopias serve to draw us forward from the present to the more desirable future.

Alternately, the discursive model of utopia is the position that ideal projections of any given society are generated by authors within it who assume and articulate critical relationships with the socio-political realities they confront. Utopia thus finds its implicit base in a rejection of certain salient features of actual lived experiences, and their replacement with ideal surrogates. The multitude of possible utopias is accounted for on this model not by outlining divergent end-states for a society, each of which addresses key social problems in an equally convincing fashion, but in making the assumption that each coordinate of moment-in-time and location-in-space will have a different counter-world that corresponds to it in a utopian relation. The discursive model's sphere of application is chiefly historical

C.C. Yorke (✉)
The Open University, Milton Keynes, UK
e-mail: christopher.yorke@hotmail.com

J.S.J. Schwartz and T. Milligan (eds.), *The Ethics of Space Exploration*,
Space and Society, DOI 10.1007/978-3-319-39827-3_5

and literary, in helping to explain why the utopias of different eras have distinctive themes and forms. Discursive utopias aid us in understanding how the undesirable present pushes us toward better futures.

The final model of utopia we will consider is the horizonal model, which posits that utopia is never reached but always aimed toward. By stipulative definition, this model places utopia on the horizon of what is (least) possible and (most) desired: so, if we are able to reach somewhere that was once on our horizon, it cannot be utopia anymore; for utopia will be whatever currently occupies that horizonal space in the distance, wherever it is that we may be positioned. Its sphere of application is chiefly philosophical, in understanding why utopia-qua-place cannot be reached but utopia-qua-concept still retains its utility. Horizonal utopias show us that our relationship with utopia is spatio-temporally static but volitionally dynamic: whether we are repulsed by the present or attracted by the future, utopia will always be off in our distance, its desiderata fluctuating as we approach it.

I will consider each of these models in turn, with specific attention given to their conceptual suitability to space travel and colonization.

5.2 The Teleological Model

Perhaps surprisingly, the teleological ('pull') model of utopia seems to be a bad fit for space colonies, or even temporary communities of space travelers, at least as we understand them today. This is because space itself is not suited for human life, or any other complex lifeform (that we have knowledge of) for that matter, and thus constitutes an inappropriate and unappealing 'end-state' for humankind. The astronaut must exercise continuous caution almost to the point of neurosis to ensure her continued survival in the highly-regulated and claustrophobic artificial environment of a space shuttle or space station. Putting a species in an environment wherein it would die if it behaved naturally cannot produce 'optimality', any more than throwing a fish onto dry land or in a fish bowl is putting it in 'utopia'.

All the same, according to the theory of evolution, we humans owe our very existence to tetrapods—shallow-water fish who were environmentally incentivized to venture onto land, pushed to change or perish. Nevertheless, it is bizarre to think of a human as a perfected fish, or of a fish as an imperfect human. Presumably, tetrapods did not venture out of the water because of a utopia they wanted to realize or a human form they recognized they could someday become, but because they were contingently forced to—fish weren't meant to become humans; things just happened to turn out that way.

The relation of fish to human could be said to mirror the relation between human and utopian, as the utopian presumably constitutes a future stage of our evolution, rather than a past one. This raises the question: are humans somehow unconsciously destined to become utopians, as fish were, in some sense, bound to become human? Or, relatedly, would it possible for humanity to arrive at utopia by accident, via the mysterious process of evolution? Most standard conceptions of utopia run counter to these kinds of notions, as utopias are generally understood as products of human intention; they do not grow like a cancer or an embryo without our knowledge and volition.

To advocate a teleological model of utopia in space, one would need to subscribe to some variety of transhumanism—the idea that our species can engineer its own evolution purposefully to fit new environments or social conditions—in order to reasonably ground the claim that the end-state of extra-planetary habitation is an appropriate goal for humans to utopianize. Transhumanism builds human intentionality into the evolutionary changes that would be required for adjustment to life in space, and might also ameliorate the concern of inhabiting outer space as being an inappropriate species-*telos*, by changing the species itself. I make the strong claim that some sort of human evolution is necessary to realize a teleological utopia, because while it is true that great strides in technology are possible which could open up more comfortable and less stressful physical environments in space, meant to mimic the desirable features of Earth, it's difficult to imagine that humans could be psychologically at ease with living in space without a radical transformation of consciousness occurring on this planet beforehand, a cultural evolution to accompany its genetic correlate. In this, I agree with William Bainbridge's assertion that "outer space is so different an environment from the one our species evolved within, that a thorough transformation of ourselves will be required before we can make it our home".[1]

Nevertheless, due to the limitless avenues of exploration and discovery that space opens up, if such adventures were open to our species it would always be premature to posit any 'final resting place'. Whether we evolve, make space conform to us, or both, history will continue to march on, and new information and challenges will confront humanity. So the teleological model ultimately fails not due to contingent limitations on terraforming, eugenics, or the lack of any other scientific 'silver bullet', but in the very fact that it aims for a singular ahistorical end-state in a near-infinite universe of possibilities.

5.3 The Discursive Model

Alternately, the discursive ('push') model of utopia offers post hoc explanations of the historical or cultural necessity of certain utopias being produced at certain times in certain places. By parsing utopian visions as the next logical step in a given discourse between a society and its citizens, or as the inevitable byproducts of a set of material conditions thrown up by historical processes, some proponents of this model may even attempt to concoct a social science of human aspiration. At the same time, the figure of the author moves to the background in this picture, a mere

[1]Bainbridge (2009, p. 521).

puppet of history, because—taking the following superficially plausible assertion for instance—'If Thomas More hadn't written *Utopia*, someone else would have.' Utopian visions are merely excrudescences of *Zeitgeist* on this view; discursive utopias can be more accurately described as 'radiated' than plotted.

Applied to the topics of space exploration and colonization, the discursive model could perhaps tell us a story outlining why so many utopian and dystopian novels were set in outer space in the early twentieth century, as well as the role of science fiction as a genre in motivating real-world attempts at space travel. But this would constitute just another historical phase in a series of such currents: it would give us the how and the why of the utopia we were pushed toward, by outlining the flaws in the society that pushed its author away. On this account, utopias set in space would be aggregated in one chapter in a longer tale of political imaginaries, to be supplanted by some other type of utopia which will only come into focus at some point in the future, with the benefit of historical hindsight.

In other words, the discursive model is better suited to help us understand the social realities of the present rather than the objective value of any given utopian projection of the future. However, while one might speculate on the precise relationship between a utopia and the society that presumably necessitated its description, this will remain a matter of interpretation rather than objective fact. So while the discursive model offers theorists an interesting tool for analyzing past utopias, it cannot grant us the clairvoyance to predict the next wave of utopian literature, nor any conceptual means by which we can sift between diverse visions of utopia for the purposes of actual suitability to the human species, in space or elsewhere.

5.4 The Horizonal Model

We are left with the horizonal ('static-dynamic') model of utopia to consider, perhaps best articulated by Ernst Bloch in his *The Principle of Hope*. Bloch describes the 'Utopian Principle' as human hope, situated on "the horizon of the consciousness… toward possibility that has still not become".[2] Hope has no final resting place, because we cannot hope for what we already have. We can only inhabit past utopias, the world(s) we desired yesterday, and so what we cannot have today becomes our new concept of utopia. The history of humankind is thus that of living through a succession of expired utopias; a condition which is static, in that a live utopia can never be reached, but also dynamic, in that no two expired utopias will be exactly alike. Only if there were no legitimate object left to hope for—a rare collective psychological state coupled with an appropriately superabundant physical state—would utopia cease to exist as a category in the horizonal sense.

[2]Bloch (1986, p. 7).

On this model, space might be conceptualized as yet another limit of possibility and desire to be eventually realized and transcended, were it not limitless for all intents and purposes. In the unconquerable vastness of space, the unyielding aspirations of humankind might meet their material match. If, indeed, space represents the final frontier of human desire, because the conditions for its complete possession can never be met, then the horizonal model of utopia may indeed have a novel and central role for outer space to play within it. Given that the other models of utopia seem to lack similar suitability, Bloch's horizonal model merits closer scrutiny. There is an ineliminably personal dimension to the experience of hope for the 'Not-Yet-Become',[3] and our more mundane desires as well. Yet, we can also easily find examples of shared hopes great and small. Alone with my test paper, I hope to pass an important examination; in the stadium bleachers, we all hope that the home team wins the game. The self-regarding hope is, in many cases, apolitical, while shared hope is political by definition. Individually, our hopes might conflict with each other's (you might, for example, want me to fail my test while I want to pass), and indeed my own set of personal hopes might suffer from internal contradictions (I might want the home team to win, but also not want them to win, if the resultant party will disrupt my studying for tomorrow's test). Due to the multifarious and multitudinous expression of possible hopes in Bloch's schema, Vincent Geoghegan writes that if "the production of utopias is a response to fundamental desires and dispositions in individuals across time and space... it is difficult to see what does *not* count as utopian".[4]

A possible defense of Bloch's massive work on hope against Geoghegan's charge of conceptual vacuity can be found in the underlying theme of harmonizing of the desires of the individual with the desires of the group, wherein the group does not tyrannize the individual, and the individual is not parasitic upon the group (exercising contrary, selfish desires at the expense of the collective), a state he calls 'homeland'. Before 'homeland' is realized, "man everywhere is still living in prehistory, indeed all and everything still stands before the creation of the world, of a right world".[5] It is only in the articulation of the group's desires that a utopia is recognized and made conceivable as a goal for coordinated political efforts toward it. Individual desire must be educated and subsumed for the common good; for, if the fate of society were left to the war of individual desire against individual desire, then it would become a nothing more than a Hobbesian state of nature. Pure utopian aspiration (as opposed to the hope of immediate strategic gains) in this scenario would only be made possible in the form of a social contract, mutually

[3]This term is Bloch's shorthand for a concrete utopia on the horizon of possibility (Bloch 1986, p. 11).

[4]Geoghegan (1996, p. 151).

[5]Bloch (1986, p. 1375). To be an 'objective hope-image', homeland must involve the coordination of all citizens' desires. It represents the cessation of conflict resultant from individuals adhering to a multitude competing subjective hope-images, and thus answers the anti-utopian criticism launched by Karl Popper in *The Open Society and Its Enemies* (1945).

circumscribing the scope of felicitous desires. Of course, if we could actually curb human hopes according to the dictates of a contract, the world would be a very different place than it is. Still, postulating a meta-utopia of harmonized desires (homeland, as opposed to mere horizon) gives Bloch a possible answer to Geoghegan, as then there would be a sorting mechanism for differentiating particular desires from universal desires (because, of course, not all particular desires are candidates for universality), and thus not all things would count as potential expressions of the utopian impulse.

Another potential weakness of this model is that the goals of groups are equally capable of conflicting with each other as the goals of individuals are. But this in itself is not fatal, insofar as one group might have appropriate goals in sight while another does not (although the ultimate means for arriving at such judgements are internally lacking on this model). More troubling is the objection that the desires of groups are not always morally defensible simply by virtue of their being the desires of groups. While desire is a necessary stimulus to human action, there are many human desires such that, from a moral perspective, the world would be a better place were they not acted on. Thus it looks like Bloch's account needs something like Immanuel Kant's categorical imperative (the idea that something is morally permissible only if the consequences of everyone doing it are acceptable) in order to sort out permissible group desires (those which can morally be universalized) from impermissible ones (those which cannot). This, or a similar conceptual fix, could serve to ameliorate these criticisms of the horizonal model of utopia.

Returning our discussion to the topic of space, let us consider whether the activity of space travel and colonization is collectively morally defensible, and thus whether it is a good candidate for Blochian utopian desire, or whether it constitutes, as a critic might phrase it, a morally irresponsible squandering of limited resources. To consider space exploration to be a pursuit of merit, value on some axis must be brought into the equation: for example, the recent discussion of the possibility of mining minerals on asteroids constitutes a potential economic defense of the space program, while the search for habitable exoplanets with an eye to eventual colonization could be said to contribute, in an as-yet abstract manner, to the ultimate good of the survival of our species. Finally, while this feature is often ignored, the continued attempts to explore space fuel a great number of intangible goods, specifically those described by William Bainbridge as 'far-out' justifications for the space program: serving as expressions of, and inspirations for, human hope, wonder, and curiosity.[6] Negatively, however, the value of space exploration becomes clearest when we imagine the consequences if it were abandoned as an aim for the human species.

[6]Bainbridge (2009, p. 518).

5.5 Possible Consequences of a Fixed Horizon for Humankind

In the barest physical sense, but also in an important psychological one, the abandonment of space exploration is the acceptance of a fixed horizon for humankind. But what would it mean for humans to lose the potential of new horizons, new frontiers? There are different ways to project this development: as a net loss or a net gain.

Trivially, humankind with a fixed horizon is stagnant, but stagnant only insofar as it has grown to the limit of its desires. Developmental satiety is an enviable variety of stagnation. On the individual level, one can see this in a person who retires with dignity: "I'd gone as far as I wanted to". As a species, we might consider that *homo sapiens* has come too far, and seen too much, to indefinitely keep pressing new boundaries. A reliable sign of maturity, after all, is the abandonment of frivolity; and the space program is easy to caricature as the height of human frivolousness or *hubris*.

Albeit stagnant on a single axis, one might yet predict exciting cultural advances to be gained from the abandonment of the space program. Perhaps a new generation who fully inhabit the present, who maturely confront the hard realities of the now, abandoning ideological comforts and diversions, will emerge. The human without any future utopia to yearn for, the human of fixed horizon, may after all be more motivated to engage actively with the given facts of today's problems.[7]

On this 'net gain' side of the debate, we can place Immanuel Kant's "To Perpetual Peace" (1795), which was predicated on the assertion that humans would only cooperate with each other and overcome their quarrelsome natures when there were no more uninhabited (contingently habitable) areas left in the world, at which point the species would be rationally compelled to accept a globally consolidated rule of law.[8] Until that projected point in human history, there would always be a place to flee to in case of conflict, and therefore insufficient motivation to choose a life guided by reason over a life guided by instinct. Kant's quasi-utopia of perpetual peace obtains only when there is nowhere left to run. Thus, if outer space in all of its indefinite extensibility is proven to be habitable, a Kantian perpetual peace in *any* world is thus unlikely to obtain on the grounds of his own argument, for as long as the possibility of perpetual flight is kept alive in the minds of humans, there would be no necessary correlative submission to the dictates of reason. Abandoning

---

[7]Provoking such a cultural awakening was Frederick Engel's key motivation for attacking utopian socialism in "Socialism: Utopian and Scientific", and for Karl Marx's condemnation of religion as the "opium of the people" in "Towards a Critique of Hegel's *Philosophy of Right*: Introduction".

[8]Kant (1983, pp. 121–125). In his words, nature "has taken care that men can live in all regions of the world... Through *war* she has driven them everywhere... Also through war she has constrained them to establish more or less legal relationships". (p. 121) His note on the '*life of the hunter*' seems to strongly confirm my reading of him that access to space for retreat strongly correlates with the continuation of warring behavior (p. 122).

the space program would greatly assuage that concern for Kant, specifically if it were somehow sabotaged in such a way that it could not be reversed or replicated. I will refer to this position, which recommends a terrestrially restricted existence for the species, as 'consolidationism'.[9]

On the 'net loss' side, we find Frederick Turner's "Frontier Thesis" (1893), a sociological statement drawn from evolutionary theory, which posits that the character of a people is determined by the severity of the obstacles they overcome. For theorists like Turner, space would represent an endless series of frontiers of the utmost severity, and thus constitute an endless repository of potential cultural developments. The failure to pursue space's exploration would constitute a missed opportunity of species-wide consequence, and may even result in cultural regression. I will refer to this position, which recommends a complete lack of spatial boundaries for the species, as 'frontierism'.

To decide whether it is frontierism or consolidationism that should inform our attitude towards the space program, we will need to consider each in turn.

5.6 Frontierism

In describing the American pioneers' work in pushing the country's borders westward, Turner wrote: "...at the frontier the environment is at first too strong for the man. He must accept the conditions which it furnishes, or perish...",[10] and further that "In this progress from savage conditions lie topics for the evolutionist".[11] His overarching claim is that the quarrelsome, innovative, and unique character of the American people was derived from their antagonistic and difficult interactions with conditions on the frontier, and that without those experiences, the nation would be weaker and lesser than it currently is.

Turner's nationalized version of evolution is more sociological than philosophical, but the problem we are tasked with addressing is a philosophical one. So it is that I am compelled to derive a toy ethical position—'frontierism'—from Turner's thesis. Frontierism is a hypothetical branch of ethics which holds that virtue is generated through the overcoming of great obstacles. It's a truism that if there were no obstacles, there could be no virtue; but the frontierist further claims that the greater the obstacles that one faces are, the morally greater one has the potential to become in the process of overcoming them (assuming, of course, that one is not crippled or killed in the attempt to overcome, which would be indicative of poor judgement or bad moral luck).

From the point of view of this toy frontierism, space is the ultimate obstacle, as it poses the most peril and challenge to human life. Therefore, the attempt to colonize

[9]I am indebted to Tony Milligan for suggesting this nomenclature.
[10]Turner (1962, p. 4).
[11]Turner (1962, p. 15).

space has the best odds of producing the greatest virtues in humankind. As utopia is the least likely outcome of human history (always on the outermost horizon of possibility), and outer space is the least likely venue for a utopia (being the least hospitable environment for sustaining human life), for frontierists it stands to reason that we can become the best possible kind of people only by attempting to achieve this (most) difficult feat: building a utopia in space. On a similar line, Linda Billings notes that "Space exploration offers tremendous opportunities to take extraordinary risks and thus it promises great challenges to the human mind and spirit. Intellectual and spiritual growth are more than worthy goals of future space exploration efforts".[12]

Since space can never be fully mapped or tamed, it constitutes an inexhaustible series of frontiers with which the development of humankind can be spurred forward. This picture fits with Bloch's horizonal model of utopia: the premise that one set of social aspirations eventually generates the social conditions from which the next, substantially refined, set of aspirations can be articulated. More succinctly put: *utopia is born from utopia*. Parallels with transhumanist theory can also be drawn, in that the indefinite improvement of society and the indefinite improvement of the species each have no conceptually requisite endpoint, and these two goals can easily be linked in tandem. Human beings might not actually be *homo sapiens* if and when a utopia in outer space were achieved, but rather some successor species, all-but-unrecognizable to their ancestors, with their own new utopia in view, which may be literally incomprehensible to us. The unlimited frontier beckons, with strange new vistas to behold, and (potentially) new eyes to view them with.

5.7 Consolidationism

Kant saw the utility of humankind's innate urge for conflict cashed out as a civilizing force, as all wilderness would eventually become cultivated by populations displaced by war, ultimately for the collective benefit of the species. As this process of physical cultivation reached its point of completion, humans would be forced by their new circumstances (having nowhere else to go and settle) to accept reason as their guiding ethical and political principle henceforth, resulting in a consolidated global socio-political peace. Consolidationism of Kant's variety models humanity as a liquid, which will come to rest peacefully once it fully spreads to the edges of its container.

This point, when the horizon disappears completely, is also the worldly death of frontierism, and it sees its parallel described by Turner at the point at which the American West had been explored to its logical endpoint, the Pacific Ocean. Turner lamented this development, observing that the exhaustion of frontier meant the beginning of an unproductive type of turmoil, rather than Kant's projected

[12] Billings (1997, p. 190).

tranquility: "This, then, is the real situation: a people composed of heterogenous materials, with diverse and conflicting ideals and social interests, having passed from the task of filling up the continent, is now thrown back upon itself, and is seeking an equilibrium".[13] Actually attaining such a geographically-bounded equilibrium would result in cultural stagnation on the frontierist view—a much more dangerous state than that of perpetual social upheaval.

Humans have, however, proven be very adept at opening up new territories, previously thought to be uninhabitable, and settling them. The icy wastes of Antarctica, remote mountaintops, the ocean floor, and the vacuum of space, have all been temporarily colonized by humans: the chief obstacle to permanent habitation of these areas seems to be pecuniary rather than conceptual. Once again, outer space poses a special case, in that it appears for all intents and purposes to be endless. It is a live conceptual possibility that, at some point in the species' future, humans could live off-planet, conceivably outside of their native solar system, whether or not this is actually practicable in terms of finance. As stated previously, this possibility threatens a Kantian perpetual peace in this and any other world, because as long as flight from conflict is possible, there is insufficient motivation to prefer a life of reason to a life of instinct, and so no necessary acceptance of the rule of global law.

Consolidationism, as a toy philosophical position which denies the possibility and desirability of novel horizons, is easy to caricature as an embrace of hopelessness, impotence, and the anti-utopianism of the status quo. However, it need not be any more than the stoic acceptance that we are better off as a species without employing futile means to attempt the achievement of Quixotic ends. For the consolidationist, the potential exploration and colonization of space only serves to distract us from real issues on this planet, where the entire population—minus less than a dozen species members—actually lives. It is only in the abandonment of the hope of escape to a counter-world that lasting peace can be established in this one. The only practical and theoretical challenges left after territorial consolidation would be those raised by constant adjustment to what is, and learning to filter out what might be. Consolidationism thus represents the ultimate triumph of (Kantian) reason over passion.

5.8 Conclusion

The claims of the consolidationist camp regarding the socially disruptive potential of space exploration for human peace seem exaggerated, and rest on a convoluted internal logic: for we cannot pretend that space exploration and colonization are not contingent possibilities when we have clear evidence that they are. Furthermore, even if space travel were only a conceptual possibility, it—or some other escapist

[13]Turner (1962, pp. 220–221).

imaginary like the "empty longing… for the *golden age*" Kant writes of[14]—would no doubt still play the role of ideological consolation, and motivate political activity along its line, unsettling the stagnant peace he predicts must obtain.

A much more promising theoretical line would be to side with the frontierists: to view our species as a fundamentally pioneering and self-reinventing one, generating virtue via facing challenges. If generally adopted, this view has the potential to culturally and politically unite humanity, in working together to enable collaborative efforts traverse the harsh conditions of endless space and eventually settle it. A horizonal model of utopia is a natural and useful socio-political correlate to this framework. Bloch's vision of utopia—as a fluid set of shared aspirations—can be adapted to fit any novel circumstances this grand endeavor may encounter, while at the same time ideologically preserving species-cohesion even as evolution pushes us ever onward in unanticipated directions.

Indeed, many of the conceptual (and traditional literary) conditions for a thriving utopia are met by the necessities of life in space: an isolated locale, a disciplined population, and an easily-identifiable historical (re)starting point. Reframing the question of politics and ethics in space along utopian rather than traditional provincial lines promises to help us avoid making the same mistakes twice, in space as well as on Earth. For if national interest continues to win out against species-interest, if space is conceptualized merely as 'Earth writ large', then we have forfeited space's considerable utopian potential.[15]

References

Bainbridge, W. S. (2009). Motivations for space exploration. *Futures, 41*, 514–522.

Billings, L. (1997). Frontier days in space: Are they over? *Space Policy, 13*, 187–190.

Bloch, E. (1986). In N. Plaice, S. Plaice, & P. Knight (Eds., & trans.), *The principle of hope.* Cambridge, MS.: The MIT Press.

Geoghegan, V. (1996). *Ernst bloch.* New York: Routledge.

Kant, I. (1983). In T. Humphre (Ed., & trans.), *Perpetual peace and other essays.* Indianapolis: Hackett Publishing Company Inc.

Turner, F. J. (1962). *The frontier in american history.* New York: Rinehart and Winston Inc.

[14]Kant, "Speculative Beginning of Human History", in Kant (1983, p. 59).

[15]In addition to the editors of this volume, I would also like to acknowledge the formative feedback of Jon Pike and Alex Barber on earlier drafts of this article.

Part II
Normative Ethics

Chapter 6
Cosmological Theories of Value: Relationalism and Connectedness as Foundations for Cosmic Creativity

Mark Lupisella

6.1 Introduction

Modern science has marched in remarkable and unsettling directions. Theories of the very big (such as general relativity and big bang cosmology) and the very small (e.g. quantum mechanics), as well as our increasing understanding of cosmic evolution and the evolution of life, have proven to be robust and powerful—not just in terms of what we've learned and what we are able to physically do in our world, but also in terms of how we experience, understand and think about ourselves in a broader context. There are seemingly endless attempts at unified theories such as quantum gravity, quantum cosmology, and string theory that challenge our minds and assumptions and presumably promise yet more profound developments. Philosophical thinking and worldviews have been shaped by modern science and vice versa, and here we will explore some of those relationships and how they may inform, or least permit, an exploration of cosmological theories of value.

Attempts to link disparate areas such as quantum mechanics, cosmology and value theory should be approached cautiously to say the least. Nevertheless, the approach here will be to draw from fields of science such as relativity, quantum theory, and cosmology, as well as from cultural evolution and philosophical domains such as value theory and process philosophy, to support the development of a relational framework that emphasizes the relational nature of the universe as a compelling reality. The strong version of this relational articulation will introduce a "connection-action principle" which leverages the connectedness of the universe to suggest how relations, action, diversity, and creativity can arise. This principle will be used to help support the possibility of intrinsic value, particularly as it relates to cosmic evolution, which can then act as a foundation for a strong theory of cosmological

M. Lupisella (✉)
NASA Goddard Space Flight Center, Greenbelt, USA
e-mail: mark.l.lupisella@nasa.gov

J.S.J. Schwartz and T. Milligan (eds.), *The Ethics of Space Exploration*,
Space and Society, DOI 10.1007/978-3-319-39827-3_6

value. We will also briefly explore two weaker theories of cosmological value that do not require conceptual formulations of the kind just noted, but will nevertheless draw from and include components of modern cosmology and relational philosophy more generally.

6.2 Cosmic Evolution

One of the most profound implications of modern science is that of an evolving universe. A primary pillar of the evolving universe was erected by Edwin Hubble when he discovered an expanding universe from the top of Mount Wilson in the 1920s. An expanding universe, along with other evidence, led to inferences of a point-like, infinitely hot and dense initial condition giving rise to an "explosion" of space-time itself. This Big Bang cosmology has been supported by many converging lines of evidence—although it is by no means the end of the story (e.g. see Smolin 1997; Steinhardt and Turok 2004). From this skeleton picture, scientists began to fill in the pieces, slowly painting a reasonably cohesive evolutionary picture of our universe.

Cosmic evolution has created a staggering myriad of phenomena and objects, including intelligence, in a long process of expansion, cooling, element composition, stellar explosions, element distribution and the emergence of self-assembling self-replicating molecules, upon which natural selection acted to develop beings such as ourselves—beings who have come to learn that we've evolved from the basic elements of the universe itself, from "star stuff", as Carl Sagan was fond of saying. We have learned that we are, in an important sense, children of the cosmos.

6.2.1 Relativity Theory

A key physical theory related to the universe is Einstein's theory of relativity. One notable distinction between Newtonian mechanics and relativity is that Newton's theory could be considered an ontological theory, while Einstein's was more of a theory of measurement (Goldberg 1984). Einstein was in some sense forced to adopt a theory of measurement in order to explain the picture he saw emerging from the state of physics at the time. One key mechanism, perhaps the sole mechanism, by which Einstein arrived at his conceptions, is his extensive consideration of simultaneity, or more fundamentally, the limitations of information communication via the finite speed of light. Einstein realized that information could not be communicated any faster than the speed of light, where Newton had assumed that information was communicated instantaneously. This realization played a critical role in the construction of the special theory of relativity and resulted in a theory that was arguably more about information and how the world is measured, as opposed to how the world *is*. Further, measurement results depended on how the observer is related to the observed—specifically, the velocity of the observer is critically tied to measurements results.

Einstein's equations of general relativity led to solutions suggesting an expanding universe. Einstein himself at first rejected this theoretical implication of his equations and committed what he later called his greatest blunder by introducing a cosmological constant which made the universe static—the prevailing metaphysical worldview in the West for over two thousand years. However, the theoretical work of Alexander Friedman and George Lamaitre, followed by the empirical work of Edwin Hubble who measured galaxies receding at ever increasing speeds, convinced most scientists that the universe was indeed dynamic and expanding from what was likely to have been a single very dense "big bang" event giving rise to the universe we see today.

6.2.2 *Quantum Theory*

Quantum Mechanics is also arguably a theory of measurement and has significant historical interpretations (e.g. the Copenhagen interpretation) that suggest the importance of the act of observation and measurement in determining specific results (Heisenberg 1958; Bub 1974; Jammer 1974; Albert 1992; Omnes 1994). Relational quantum mechanics (RQM) is a relatively recent interpretation of quantum theory that suggests state and physical quantities refer only to the interactions, or the relations, between systems. RQM sees quantum theory as representing the full web of relations and interactions between all physical systems, rendering the state of an isolated system meaningless (Rovelli 1996; Van Fraassen 2010).

RQM is an important interpretation for the treatment of this essay because importantly, RQM appears to be a very general interpretation in the sense that all physical interactions are subject to a relative specification irrespective of the nature of the objects involved (e.g. conscious observers do not constitute a special frame of reference). Any typical physical interaction is the functional equivalent of a measurement. In this sense RQM can be seen as a more general form of the relative state formulation of other interpretations of quantum mechanics. It goes well beyond the blending of observer, measuring apparatus, and object and essentially replaces observations with all other relative physical systems as the relevant frame(s) of reference.

6.2.3 *Complexity and Self-organization: A Self-organizing Universe?*

There has been an increasing recognition of the possibility that the universe has been steadily evolving toward increasing levels of "complexity" (Chaisson 2009; Christian 2009)—despite notorious difficulties in defining exactly what complexity is. One simple definition of complexity for the purposes of this essay is "the study of the phenomena which emerge from a collection of interacting objects" (Johnson 2009, 3).

Regarding the universe as a whole, Paul Davies writes: "I believe that there's a sort of 'law of increasing organized complexity' operating in the universe. It's not quite a law in the same sense as, say, Newton's law of gravity, more like a *tendency* or *trend* (italics added), but its manifestation seems unmistakable. There really does seem to be a general tendency in nature for increasing organizational complexity (or depth)." (Davies 1995, 105).

If a potential trend or law of increasing complexity and/or self-organization for the universe as a whole is going too far, then we might at least consider increasing complexity as a trend for biological and cultural evolution, and possibly even its ethical relevance (Smith 2014). Paul Davies notes that life and consciousness can be seen as *emergent* properties, as "...something that emerges in a physical system when it reaches a certain level of complexity" (Davies 1995, 98). A similar idea is suggested by Stuart Kauffman to explain the origin of life emerging as a phase transition (1995). An emergent property is more about the relationships, the functional organization, or perhaps information content of any given system, as opposed to what it is made of.

6.2.4 Cultural Evolution

Biological evolution, possibly along with other phenomena having to do with complexity, emergence, group selection, and human socio-cultural dynamics, has given rise to powerful forms of cultural evolution. It isn't just that cultural beings learn, share knowledge, live reasonable well in groups and exercise what is powerful and often dangerous ingenuity, it is also that humans collectively debate, pursue and achieve highly complex goals and aspirations—often with increasing levels of personal and collective awareness. Part of this awareness comes from philosophy and science—including psychology and neuroscience which are shedding powerful lights on the workings of human minds and group dynamics.

While much, and possibly most, cultural evolution is rooted in biological evolution, it is increasingly defensible to think that human behavior is showing signs of transcending our selfish genetic heritage, and that trend may be applicable to extraterrestrial forms of intelligence if it exists (Lupisella 2013). Some of these signs come from richly complex intellectual explorations of what human beings aspire to, personally and collectively. Such philosophical explorations have helped humanity forge a unique power to realize what otherwise might be unachievable aspirational pursuits—sometimes carefully and wisely, sometimes not. Regardless, our careful contemplation, debates, and implementations regarding what is valuable and why, form a cornerstone of the modern human condition.

Our highly adaptive and flexible brains have allowed a contemplative and operational capability to evolve and create a compelling form of cultural evolution, i.e. normative aspiration. Normative aspiration exerts tremendous influence on human beings and the rest of our world and is likely to continue to do so with ever increasing power. If there isn't a law of cosmic self-organization, then there may at least be a kind of law of increasing complexity and/or organization in cultural evolution. How this influence is used, how it is managed and pursued, often rests largely on what is valuable and why—i.e. on our normative aspirations.

6.3 Value Theory

As we head for our brief treatment of relationalism and how it may inform cosmological theories of value, it will be helpful to touch on a few relevant areas of philosophical value theory. Two key interrelated areas of value theory are issues surrounding facts vs. values and issues regarding instrumental versus intrinsic value (Moore 2004, explores these issues explicitly in the context of relationalism). Also, a kind of meta-issue worth touching on briefly is the psychology of philosophy.

6.3.1 Is-Ought and the Naturalistic Fallacy

There is a long-standing distinction in value theory between facts and values. Related to that distinction is an observation first emphasized by David Hume, that values (or "oughts") cannot be strictly derived by something so different as facts, or the way the world "is". Something similar, the "naturalistic fallacy", also cautions against making conclusions about what is good based on natural properties (Moore 1903). It is prudent to make note of this consideration because an underlying theme in this treatment is that something about the way the world is can inform our perceptions of value and our aspirations. If we so choose, aspirations about our world can be very much informed by what we learn about the world, but do not necessarily need to be constrained by such perceptions of reality or "facts". It would not be unreasonable to constrain conclusions about value and ethics to what we think are real facts about the world—it's an understandable choice for a variety of reasons. However, similarly, it is not unreasonable to use perceptions of reality to significantly inform our explorations of values and ethics while not necessarily being constrained solely by such facts—as the naturalistic fallacy cautions against (Putnam 2002).

6.3.2 Instrumental and Intrinsic Value

Another notable issue in value theory is to address if or how to "locate" value(s) and their source(s),[1] and from that to at least partially derive some rationally

[1]Odel (1980) notes that when fact is interpreted to "embrace the total network of truth, from which it passes readily to encompassing the whole of reality", the fact-value problem is "almost converted into 'Where in the world are we to locate value?'" p. 4.

justifiable value(s) to whatever extent possible. Viewed broadly, two themes feature strongly in value theory and ethics: deontology and teleology. A simplified but useful way to distinguish between the two is that deontology generally emphasizes the value of acts, whereas teleology emphasizes the end or consequences of acts (Nozick 1981; Pojman 1995). Both approaches appear to ultimately involve some appeal to basic or fundamental forms of value. Teleological views, while focusing on consequences, must ultimately give some account of why certain consequences or ends are preferred or valued over others. Some form of fundamental value, perhaps something akin to intrinsic value, can be helpful regardless of which approach is adopted.

But truly *intrinsic* value—i.e. value that is truly independent of valuing agents—seems difficult to justify (Smith 2009). What does seem reasonable to defend is that value can often be understood at least the relational sense that value is often associated with situations where something is of value to something else. The relationship of value between "entities" is what gives rise to "instrumental" value. We know things are valuable to human ends as a function of how things are related to human beings and our interests—and as such those things at least have instrumental value for humans.

6.3.3 Psychology of Philosophy

Much of what is bound up in considerations of facts vs. values, the naturalistic fallacy and intrinsic value, and arguably philosophy in general, are factors in human psychology (Bartlett 1989). It may be informative to very briefly turn the lens of psychology—perhaps initially with an emphasis on evolutionary psychology—on philosophical explorations. Human tendencies to over-simplify and find patterns, meanings and purposes, including those that are aesthetically appealing, is infamously problematic [sometimes referred to as "promiscuous teleology" or even tenacious teleology (Kelemen et al. 2012)]. Such tendencies are arguably evolutionarily driven—perhaps most related to social status and group selection—much of which could presumably operate subconsciously and without careful reasoning or commitments to truth value given the blind mechanism of natural selection that primarily finds strategies for genetic replication. Such tendencies have many advantages and disadvantages of course, including how easy it is to mislead ourselves for what could otherwise be evolutionarily advantageous reasons—something that most brands of philosophy and science are supposed to work hard to avoid. Pursuits of grand philosophical "theories of everything", particularly those attempting to link disparate areas of science with values, should be particularly sensitive to potentially problematic psychological motivations. Nevertheless I will plunge headlong into the abyss of pattern and meaning seeking!

6.4 Relationalism

Our brief forays into relativity, quantum mechanics, and cosmic evolution, as well as some aspects of value theory, can be interpreted to suggest that something having to do with relations is important. Here we will briefly develop a broad and multi-faceted "relational framework", or philosophical relational theory, that could also be considered a relational worldview of sorts. It will serve as a context and conceptual foundational for various forms of cosmological theories of value.

There is a long history of what can be referred to as "relational philosophy", or in the indomitable spirit of "ism-izing", what might be called *relationalism* in the broadest sense—as opposed to, for example, a narrower use in "ontic-relationalism" (a relational theory of being developed by Kaipayil 2009), or *relationism* which tends to focus on specific relations themselves. Going back perhaps to the ancient Greeks and forms of Eastern thought such as Buddhism, there is a broad tradition of thought that emphasizes relations. It will be difficult to offer concise and complete definitions of a relational theory here, or what it means to engage in relational thinking in a strict sense, but as a starting point, relational thinking can be thought of as emphasizing the critical importance of relations in knowledge, reality, and in our experiences in the broadest sense. This would generally involve an emphasis on relations above other conceptualizations such as substantive "things" or objects (e.g. atoms). So in this context, relationalism is not just about discerning and studying specific relationships per se—as is done in so many domains of knowledge. It is instead more of a "meta-view" (as shown in Fig. 6.1), a broader

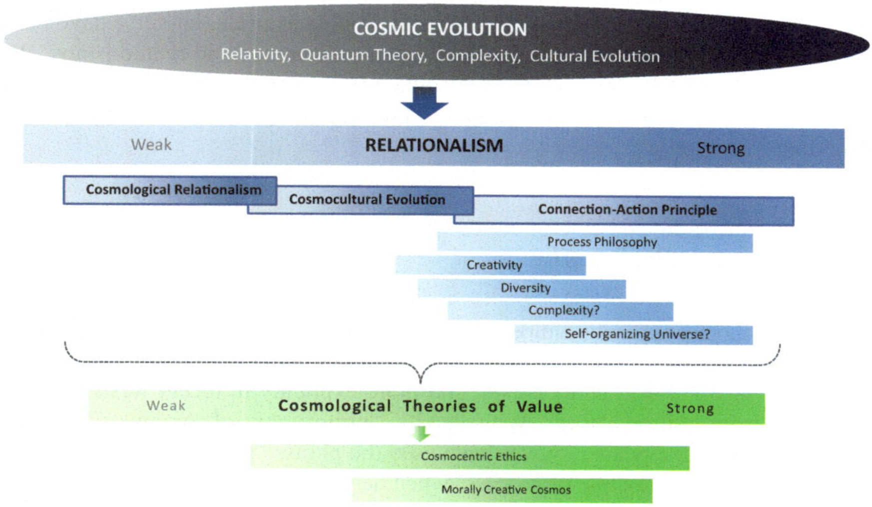

Fig. 6.1 Relational framework for cosmological theories of value

conceptual umbrella that emphasizes a critical role for relations in general—perhaps above all else (particularly in stronger versions).

In the broadest sense then, we can think of relationalism as a catch all term covering (a) relational epistemology, (b) relational metaphysics/ontology, (e.g. see Oliver 1981; Wildman 2010; Young 2011), and (c) "relational experience" construed broadly, covering other aspects of life such as relations with self and others (Gergen 2009). This treatment will emphasize concepts primarily in a and b above.

6.4.1 Relationalism in Science

At the very least, it appears that some form of relational epistemology is warranted based on perspectives of relativity and quantum mechanics noted previously. As we saw, both relativity and quantum theory suggest an important relationship between the observer and observed and both essentially eschew the idea of absolute frames of reference. Relational quantum mechanics (RQM) goes further and removes any privileged role for an observer per se and simply suggests that what is knowable and represented by quantum states is only relations and interactions in a system. Any object or system is knowable only in the way it is related to some other object or system which can act as a frame of reference, including conscious observers as well. Rovelli (1997) takes RQM even further and explores a potential connection between quantum theory relationalism and the spatiotemporal relationalism of general relativity, again suggesting something fundamental about relations more generally.

Cosmic evolution, stemming in part from the dynamic expanding universe predicted by general relativity, also suggests a certain kind of what might be called "weak relationalism" in the sense that we, and all objects in the universe have been created by an unfolding, an emergence of cosmic dynamics and relationships. We are importantly related not only to fundamental physical reality, but we are critically related and dependent on the dynamics and unfolding of that reality, on its evolution.

6.4.2 Process Philosophy

Perhaps the most prominent philosophically rigorous form of relational thinking is "process philosophy" which arguably has its first articulations in ancient eastern views such as Buddhism and Daoism, and more explicitly in recent western views stemming from Alfred North Whitehead's "philosophy of organism". While Whitehead himself appears to never have used the phrase "process philosophy" per se, he did invoke a 'principle of process' when he wrote of an entity that "Its 'being' is constituted by its 'becoming.' This is the 'principle of process.'" (Whitehead

1929, 1978, p. 23).[2] Notably, Whitehead does not prioritize being over becoming (Kraus 1998).

Of particular note for the purposes of this essay is Whitehead's emphasis on 'creativity' as the ultimate reality, the ultimate source of all that is. Indeed, it appears that Whitehead coined the term 'creativity' (Meyer 2005). Creativity is the most general and most fundamental source and category of reality and all stems from it, even God, and novelty is a key manifestation of creativity. (For more on Whitehead and process philosophy, see Lucas 1989 and Mesle 2008). Process philosophy has given rise to many wide-ranging articulations, including key distinctions between teleological views (e.g. including but not limited to theological views) and more "secular" or naturalistic views that suggest a lack of directionality toward a specific outcome (Rescher 1996). This essay will propose theories of value that lie somewhere on a spectrum between those polar views.

6.5 Cosmological Theories of Value

The broad conceptual framework for our cosmological theories of value is relationalism, which was briefly explored previously. Prior to that we touched on various aspects of cosmic evolution, including some fundamental physics and evolutionary dynamics, for which the role of relations and relational thinking was emphasized. Here will develop weaker and stronger forms of cosmological theories of value, derived partly from that relationalist thinking—which may also help justify relationalism more generally.

6.5.1 Cosmological Relationalism

The brand of "cosmological relationalism" intended here can be seen to have at least two major components: epistemological and scientific. Epistemological relationalism seems to be clearly implied by the modern physics of relativity theory and quantum theory. Many, if not most, interpretations of what those theories imply about how we know the world suggest that it is only knowable if we acknowledge the relationship of the observer and observed, that measurement results are relative, and in the case of RQM, that fundamentally, all that can be said about the world is how quantum states relate to other states.

What might this mean for a theory of value, particularly in the context of cosmic evolution? Our epistemic dependence on relations can be interpreted to imply at least a certain amount of caution and humility for the ways in which we think we

---

[2]Interestingly, the subtitle to Whitehead's (1929) book, *Process and Reality*, is '*An Essay in Cosmology*'.

understand the world and our attempts to control it. Further, what may ultimately be most valuable are relationships and interactions, in contrast to the "things" of philosophical materialism.

The science of cosmic evolution tells us that we are intimately bound up with the cosmos at least in the sense that we are a product of cosmic evolution. The recognition of our cosmic origins, along with the emerging myriad of scientific details of cosmic evolution, can imply that we might in some sense revere the universe, it's overall evolution, and perhaps cosmological processes in the most general sense (Dick 2000; Swimme and Berry 1992).

Cosmological relationalism is a weak form of relationalism and weak form of a cosmological theory of value in the sense that it is primarily a recognition of our intimate relationship to our world through the lens of contemporary science. It doesn't require any significant philosophical leap other than to plow through the naturalistic fallacy and choose to value our universe and cosmic evolution based on what science has taught us. This is a theory of value that is relatively non-prescriptive, very open-ended and realized primarily by valuing agents that are sufficiently self-aware beings that have the ability to learn about their universe and contemplate what it may mean.

Consistent with an open-ended interpretation of this kind of value, this minimal value simply acknowledges the universe as our source, as part of our history (Christian 2004)—including allowing for the recognition that the universe is not exactly made for us today in that it is not consistent with our physical health if we leave the cocoon of our biosphere. In that sense, the universe can also be seen as "hostile". But cosmological relationalism does not require a blind reverence or blind commitment to cosmic evolution or the universe as a whole—especially at the expense of other interests and commitments we may have. The "weakness" in cosmological relationalism as framed here enables us to choose to value our broader relationship to the universe, and potentially "derive" many forms of value and ethical positions from that, while at the same time avoiding problematic implications when prioritizing values and developing ethical positions.

6.5.2 Cosmocultural Evolution

As noted above, the simple open-ended idea of cosmological relationalism can be extended in a number of ways. The idea that we are importantly related to the universe is well understood by many but can be extended to think not only in terms of how beings are intimately bound up in cosmic evolution, but also how beings may have some significance in the overall evolution of the universe. Cultural evolution may have a notable significance or role in cosmic evolution, either now or perhaps more likely in the future. We might call this cosmocultural evolution (Lupisella 2009a, 2015) to capture the idea that culture and cosmos either are presently, or will be in the future, co-evolving to the extent that cultural evolution is in some sense "on par" with physical cosmic evolution—or in stronger versions, that cultural evolution could ultimately be the most influential dynamic in cosmic evolution.

Cosmocultural evolution is stronger than cosmological relationalism in the sense that cultural beings have some notable significance and possible influence in the universe. Value can be seen to be shifted slightly away from the physical universe and traditional cosmic evolution and more toward cultural beings and their evolution. On stronger versions, this can be interpreted to suggest that beings may have some degree of obligation or duty to their larger universe and its evolution. So cosmocultural evolution is more about our future relationship to the universe as a whole and choices we may make—or "obligations" we may have (in stronger versions)—to advance the interests of beings within a larger cosmic web as part of a broader balancing act with the wider universe and all other beings that may co-inhabit the cosmos with us.

This kind of view arguably lies somewhere between the extremes of teleology and non-teleology (although perhaps closer to the non-teleological end) in the sense that if there hasn't previously been meaning, purpose or direction in the universe, there is now in the emergence of highly intentional and highly aware valuing agents that may already be instantiating some directionality in cosmic evolution, and which may more strongly create cosmic evolutionary trends in the future. The universe may not have ever been intrinsically teleological, but it may be bootstrapping itself into the realm of value, meaning, and purpose through beings such as ourselves—through an emerging process of cosmocultural evolution. So some kind of directionality may now be possible and perhaps already underway. The directionality would presumably be very broad, diverse, and ever-changing, and any specific outcome would in no way be pre-determined. No specific trend would be intrinsic to the fundamental nature of the universe—it is something that would emerge via the complex and dynamic intentionality of deliberate valuing agents.

6.5.3 The Connection-Action Principle

The two views just explored can both be characterized as "weak" cosmological theories of value in the sense that they do not make strong metaphysical or ontological claims regarding a source of value in a cosmic context. The ideas articulated in this section can be considered to constitute a much stronger cosmological theory of value—a theory of value that suggests there may be a kind of intrinsic value—perhaps even "purposeful" value—at work in the universe for fundamental conceptual reasons that are arguably consistent with some scientific observations and other philosophical ideas such as relationalism explored previously.

A key notion touched on previously, as part of Whitehead's original articulation of process philosophy, is creativity. The science of cosmic evolution seems to point to a highly creative—and perhaps increasingly creative—evolving universe. We might ask why the universe is so creative? We can certainly appeal to a kind of brute fact explanation noting simply that the "laws" of physics and/or initial conditions of the universe permit it to be so. But we can then of course ask, as many have, why these laws? Why those initial conditions? There have been intriguing and compelling suggestions ranging from: (1) design (normally part of theological views) to (2) anthropic principles (e.g. Barrow and Tipler 1986), to (3) an eternally oscillating universe (going back to the Greeks and forms of eastern worldviews such as Hinduism and now by some in modern cosmology (e.g. Steinhardt and Turok 2004) to (4) cosmological natural selection (Smolin 1997) to (5) even more provocative versions of the anthropic principle that suggest conscious beings in some sense create the universe and its laws via extreme interpretations of quantum theory (Von Neumann 1932; Wheeler 1990; Davies 2009). Ultimately however, regardless of the kind of explanation for our particular laws and initial conditions, they all seem to rest on, or at least assume that our universe is dynamic—that it is a universe of *action*.

We can then wonder why there is action. Again we can appeal to a kind of brute fact explanation which might suggest that the laws of physics cause action, but as just suggested, presumably there can be action without particular laws per se.[3] An alternative speculative way to think about the source of action is to appeal to an ancient Eastern and more modern emerging Western view leveraging to oft-cited "connectedness" of the universe. The universe as a connected unity is a theme often embraced by many western thinkers now as evidenced by interpretations of quantum mechanics noted previously as well as explicitly "holistic" views (e.g. Bohm 1980; Bohm and Hiley 1993), environmental movements (Leopold 1966), and cosmic evolution more generally (Sciama 1959).

It is difficult to precisely articulate a definition of connectedness, but a common sense notion, along with some willingness to accept some unspecified details of the property of connectedness, might nevertheless suffice. Minimally, connectedness would seem to imply some kind of relationship. If there is a connection of some kind, regardless of what kind, it seems a relationship of some form would be required to make that connection meaningful and relevant, to realize and instantiate it—to make it "actionable". In fact, we may be able to use connectedness and relationality almost interchangeably since they are so close in this view. However, connectedness might be better seen as a somewhat broader, more fundamental, or slightly different construct for which relationality is one of any potential number of forms of its manifestation. Regardless, the bottom-line suggestion is that connectedness is realized through relations.

Further, for a relation to be realized or instantiated, it could be argued that something must occur, a dynamic event must manifest a relation or relationships—and this seems consistent with most forms of process philosophy. All of what we know about the universe suggests that even the most "static" phenomena are not truly static. Could this be because relationality, in manifesting connectedness, gives rise to some kind of action? Connectedness giving rise to action (or process) might

[3]In fact, the very early universe may have been highly chaotic, without much, if any, law-like behavior (Linde 1986).

be called the "connection-action principle" (CAP). Perhaps a minimalist version of the connection-action principle, put into the parlance of philosophical "property talk" would be: *The universe's property of connectedness is instantiated by action.*

A stronger version of the connection-action principle can be interpreted to suggest that the universe's property of connectedness might be realizable to *ever-increasing* degrees of action (if we allow for the possibility that possessing a property can lead to increasing degrees of its instantiation), implying a cosmic evolution toward increasing degrees of action. Increasing degrees of action could be realized in many ways, including increasing creativity, increasing diversity, increasing complexity, and increasing degrees of self-organization. The more the universe acts, the more it creates, the more new, and perhaps more robust (e.g. more complex) connections and relations that emerge, the more the universe's property of connectedness is realized—the more its potential is realized. This arguably implies a weak or soft form of teleology, suggesting the universe is actualizing connectedness and that in so doing it is evolving in many highly "active" ways, always creating, always bringing forth new connections, new relations, increasing novelty in the universe as suggested by certain forms of process philosophy and perhaps even the "temporalized" version of principle of plenitude (Lovejoy 1971). So a strong version of the connection-action principle could suggest that connectedness is increasingly realized through increasing levels action and that the greater the action (again, e.g. interactions, intra-actions, creativity, diversity, complexity, self-organization, etc.) the greater the degree to which the universe's nature, or "potential" is realized.

The connection-action principle then is consistent with process philosophy in that it suggests a conceptual mechanism that gives rise to universal creativity and process more generally. An even stronger version of the connection-action principle can be interpreted to make an ontological claim that relations are all that really exists—perhaps consistent with a relational metaphysics of sorts (Oliver 1981). Interpretations of this kind can be seen to support the idea that many conceptions of material things, (e.g. philosophical substantivism) may be a kind of illusion. Regardless of the ontological commitments, what is valuable on this view is not just relations per se, but more generally realizing connectedness through action in whatever myriad forms that may take—many of which are presumably yet to be understood or realized. What is valuable in the context of CAP is an unending unfolding of creativity, perhaps leading to ever-increasing degrees of creativity. Can we say then that ever-increasing degrees of connectedness, relations, and action represent a form, or forms, or expression of intrinsic value if it is intrinsically part of the universe realizing its nature and its potential to ever-increasing degrees?

The property of the connectedness of the universe can also be thought of as "relational potential", and interactions [and/or "intra-actions" (Barad 2007)] are a manifestation of that potential. In this sense relationalism is foundational in that connectedness is relational, and actions are relational. The connection-action principle, strictly speaking, can be interpreted to be agnostic regarding whether there are interactions among material objects (materialism/substantivism) or intra-actions within a highly blended web of relations that may not require any objects at all. However, the interpretation that only relations are ultimately needed for knowledge, combined with an interpretation that material objects do not appear to be necessarily required, suggests that the most parsimonious interpretation of the connection-action principle is that it ultimately suggests a form of *intra*-action as opposed to inter-action among discrete entities.

6.5.4 Potential Implications

This short essay cannot thoroughly explore all the potential practical and ethical implications of cosmological theories of value.[4] However, we can briefly sketch some broad outlines. For example, appealing to the universe (or possibly "multiverse") and cosmic evolution more specifically, allows us to get closer, if not achieve a link to what may ultimately be absolute (perhaps even eternal, i.e. the universe or multiverse) and at least partially understandable through methods of science, logic, and creative philosophical exploration (Vidal 2014). Related to that, theories of value that relate to the universe have the potential to provide a common frame of reference with extraterrestrial beings (Lupisella 2011, 2013, 2015). For example, a "cosmocentric ethic", that might see the universe as a priority in a value system (for which there can be weaker and stronger forms), can help establish a common value frame of reference with other beings who also inhabit our universe, including for interactions with what may be more "primitive" organisms such as extraterrestrial microbes (Lupisella and Logsdon 1997; Lupisella 2009b). Further, for both cases where sufficiently aware and capable beings inhabit our universe, or for the case where we are essentially "alone", cosmological theories of value can still provide a broad and potentially compelling sense of meaning and purpose that can guide intelligent cultural beings to work toward a morally creative cosmos (Lupisella 2014).

6.6 Summary

We have drawn briefly from relativity, quantum theory, cosmology, and cultural evolution, as well as areas of value theory and process philosophy to help develop a broad framework of relationalism, all of which has informed the articulation of three cosmological theories of value: (1) cosmological relationalism, (2) cosmocultural evolution, and (3) the connection-action principle. Cosmological relationalism is minimalist and very open-ended in the sense that it only emphasizes our scientific relationship to the universe as we understand it today—primarily that we

[4]A more through exploration of the implications of these views is forthcoming in a future book by this author (presently under contract).

are products of cosmic evolution which is a highly dynamic and relation-laden process—and suggests that a deep kind of value for the cosmos is a reasonable philosophical implication.

Cosmocultural evolution goes further and suggests that cosmos and culture are, or will be, co-evolving to the extent that cultural beings, or sufficiently valuing agents more generally, are likely to have both physical and philosophical significance for the overall evolution of the universe—possibly even obligations in stronger versions. There are very general prescriptive implications here, but choices lie primarily with cultural valuing beings to shape the future in an open-ended and morally creative manner, bearing in mind the relationality of the world and our relationships to the cosmos and all therein.

The connection-action principle goes much further and suggests that the universe's fundamental property of *connectedness* is manifested as relations and action, and hence ultimately as creativity—potentially in ever-increasing degrees. On this view, the universe's intrinsic potential, and hence perhaps then its intrinsic value, is realized through ever-increasing degrees of relationality, inter- or intra-action, and creativity—for which there can be endless myriad forms.

References

Albert, D. (1992). *Quantum mechanics and experience*. Cambridge MA: Harvard University Press.

Barad, K. (2007). *Meeting the universe halfway: Quantum physics and the entanglement of matter and meaning*. Durham and London: Duke Universe Press.

Barrow, J. D., & Tipler, F. J. (1986). The Anthropic Cosmological Principle. Oxford University Press.

Bartlett, S. J. (1989). Psychological underpinnings of philosophy. *Metaphilosophy, 20*, 295–305.

Bohm, D. (1980). *Wholeness and the implicate order*. London: Routledge & Keegan Paul.

Bohm, D., & Hiley, B. J. (1993). *Undivided universe: An ontological interpretation of quantum theory*. New York: Routledge.

Bub, J. (1974). *The interpretation of quantum mechanics*. Boston: D. Reidel Publishing Company.

Chaisson, E. (2009). Cosmic evolution: The state of the science. In: S. J. Dick & M. L. Lupisella (Eds.), *Cosmos and culture: Cultural evolution in a cosmic context* (pp. 321–259). NASA SP-2009-4802. Retreived February 2, 2016, from http://history.nasa.gov/SP-4802.pdf

Christian, D. (2004). *Maps of time*. Berkeley: University of California Press.

Christian, D. (2009). History and science after the chronomatic revolution. In S. J. Dick & M. L. Lupisella (Eds.), *Cosmos and culture: Cultural evolution in a cosmic context* (pp. 321–259). NASA SP-2009-4802. Retreived February 2, 2016, from http://history.nasa.gov/SP-4802.pdf

Davies, P. (1995). *Are we alone?*. New York: HarperCollins.

Davies, P. (2009). Life, mind, and culture as fundamental properties of the universe. In S. J. Dick & M. L. Lupisella (Eds.), *Cosmos and culture: Cultural evolution in a cosmic context* (pp. 321–259). NASA SP-2009-4802. Retreived February 2, 2016, from http://history.nasa.gov/SP-4802.pdf

Dick, S. J. (2000). Cosmotheology: Theological implications of the new universe. In Steven J. Dick (Ed.), *Many worlds: The new universe, extraterrestrial life, and the theological implications* (pp. 191–210). Philadelphia: Templeton Foundation Press.

Gergen, K. J. (2009). *Relational being: Beyond self and community*. New York, NY: Oxford University Press.
Goldberg, S. (1984). *Understanding relativity: Origin and impact of a scientific revolution*. Cambridge, MA: Birkhauser Verlag.
Heisenberg, W. (1958). *Physics and philosophy: The Revolution in Modern Science*. London: George Allen and Unwin LTD.
Jammer, M. (1974). *The philosophy of quantum mechanics*. New York: Wiley.
Johnson, N. F. (2009). *Simply complexity: A clear guide to complexity theory*. London: Oneworld Publications.
Kaipayil, J. (2009). *Relationalism: A theory of being*. Bangalore: JIP Publications.
Kauffman, S. (1995). *At home in the universe*. New York: Oxford University Press.
Kelemen, D., Rottman, J., & Seston, R. (2012, October 2015). Professional physical scientists display tenacious teleological tendencies: Purpose-based reasoning as a cognitive default. *Journal of Experimental Psychology: General*, 142, 1074.
Kraus, E. (1998). *The metaphysics of experience: A companion to whitehead's process and reality*. The Bronx: Fordham University Press.
Leopold, A. (1966). *A sand county almanac*. New York: Oxford University Press.
Linde, A. (1986). Eternally existing self-reproducing chaotic inflationary universe. *Physics Letters B*, 175(4).
Lovejoy, A. O. (1971). *The great chain of being*. Harvard University Press: Cambridge.
Lucas, G. (1989). *The rehabilitation of whitehead: An analytic and historical assessment of process philosophy*. Albany: State University of New York.
Lupisella, M. (2009a). Cosmocultural evolution: The coevolution of cosmos and culture and the creation of cosmic value. In S. J. Dick & M. L. Lupisella (Eds.), *Cosmos and culture: Cultural evolution in a cosmic context*. NASA SP-2009-4802, pp. 321–259. Retreived February 2, 2016, from http://history.nasa.gov/SP-4802.pdf
Lupisella, M. (2009b). The search for extraterrestrial life: Epistemology, ethics, and worldviews. In C. Bertka (Ed.), *Exploring the origin, extent, and future of life: Philosophical, ethical and theological perspectives* (pp. 186–204). Cambridge: Cambridge University Press.
Lupisella, M. (2011). Pragmatism, cosmocentrism, and proportional consultation for communication with extraterrestrial intelligence. In D. A. Vakoch (Ed.), *Communication with extraterrestrial intelligence (CETI)* (pp. 319–331). Albany: State University of NewYork Press.
Lupisella, M. (2013). Caring capacity and cosmocultural evolution: Potential mechanisms for advanced altruism. In Douglas A. Vakoch (Ed.), *Extraterrestrial altruism*. Berlin: Springer.
Lupisella, M. (2014). Cosmocultural evolution: Cosmic motivation for interstellar travel? *Journal of the British Interplanetary Society*, 67(5).
Lupisella, M. (2015). Life, intelligence, and the pursuit of value in cosmic evolution. In Steven J. Dick (Ed.), *The impact of discovering life beyond earth*. London: Cambridge University Press.
Lupisella, M., & Logsdon, J. (1997). Do we need a cosmocentric ethic?' Paper IAA-97-IAA.9.2.09, presented at the *48th International Astronautical Congress, Turin, Italy, October 6–10*.
Mesle, C. R. (2008). *Process-relational philosophy: An introduction to alfred north whitehead*. West Conshohocken: Templeton Foundation Press.
Meyer, S. (2005). Introduction: Whitehead now. *Configurations, 1*(13), 1–33.
Moore, G. E. (1903). *Principia ethica*. London: Cambridge University Press.
Moore, A. D. (2004). Values, objectivity, and relationalism. *The Journal of Value Inquiry, 38*, 75–90.
Nozick, R. (1981). *Philosophical explanations*. Cambridge: Harvard University Press.
Odel, A. (1980). *Exploring fact and value*. London: Transaction Books.
Oliver, H. (1981). *A relational metaphysic*. Boston: Matinus Nujhoff Publishers.
Omnes, R. (1994). *The interpretation of quantum mechanics*. Princeton: Princeton University Press.

Pojman, L. P. (1995). *Ethical theory: Classical and contemporary readings*. New York: The Wadsworth Publishing Company.

Putnam, H. (2002). *The collapse of the fact/value dichotomy*. Cambridge MA: Harvard University Press.

Rescher, N. (1996). *Process metaphysics: An introduction to process philosophy*. New York: SUNY Press.

Rovelli, C. (1996). Relational quantum mechanics. *International Journal of Theoretical Physics, 35*, 1637–1678.

Rovelli, C. (1997). Half way through the woods. In J. Earman & J. D. Norton (Eds.), *The cosmos of science*. Pittsburgh: University of Pittsburgh Press.

Sciama, D. W. (1959). *The unity of the universe*. Garden City: Doubleday & Company.

Smith, K. C. (2009). The trouble with intrinsic value: An ethical primer for astrobiology. In C. Bertka (Ed.), *Exploring the origin, extent, and future of life: Philosophical, ethical, and theological perspectives*. Cambridge: Cambridge University Press.

Smith, K. C. (2014). Manifest complexity: A foundational ethic for astrobiology? *Space Policy, 30*, 209–214.

Smolin, L. (1997). *The life of the cosmos*. New York: Oxford University Press.

Steinhardt, P. J. & Turok, N. (2004). The cyclic model simplified. *New Astronomy Reviews, 49*(2–6), 43–57. arXiv:astro-ph/0404480. Bibcode:2005NewAR..49…43S. doi:10.1016/j.newar.2005.01.003

Swimme, B., & Berry, T. (1992). *The universe story*. New York: Harper Collins.

Van Fraassen, B. (2010). Rovelli's world. *Foundations of Physics, 40*, 390–417.

Vidal, C. (2014). *The beginning and the end: The meaning of life in a cosmological perspective*. New York: Springer.

Von Neumann, J. (1932/1955). *Mathematical foundations of quantum mechanics* (R. T. Beyer, Trans). Princeton: Princeton University Press.

Wheeler, J. A. (1990). Information, physics, quantum: The search for links. In W. Zurek (Ed.), *Complexity, entropy, and the physics of information*. Redwood City, California: Addison-Wesley.

Whitehead, A. N. (1929). *Process and reality: An essay in cosmology*. 1978 corrected edition: D. R. Griffin & D. W. Sherburne (Eds.), New York City: Free Press.

Wildman, W. J. (2010). An introduction to relational ontology. In: J. Polkinghorne (Ed.), *The trinity and the entangled world: Relationality in physical science and theology*. Grand Rapids MI & Cambridge: William B. Eerdmans Publishing Company.

Young, M. (2011). Relevance and relationalism. *International Journal for Ontology and Metaphysics, 12*, 19–30.

Chapter 7
On the Methodology of Space Ethics

James S.J. Schwartz

7.1 Introduction

This chapter addresses space's impact on the methodology of normative and applied ethics. Thus it is as much an instance of metaethics as it is of other subdisciplines of ethics. My focus will be on two questions related to the activities of ethical theory construction and evaluation: First, should we have a preference for either monist or pluralist conceptions of *value* for the ethics of space? By this I mean: Should a focal point of theorizing be the identification and description of a fundamental, unified (i.e., a *monistic*) conception of intrinsic value from which our various duties can be derived? Or should theorizing tolerate a *plurality* of potentially irreducible values, each of which we are duty bound to protect? This dispute is especially salient for space ethics, given that the space environment is largely devoid of the kinds of things that mainstream, monist views typically regard as intrinsically valuable (such as the experience of pleasure under utilitarianism, or rational beings under Kantianism). Second, should a focal point of theorizing be the identification and description of a conception of intrinsic value in the first place? There are, after all, normative theories such as virtue ethics that attempt to directly circumscribe the features of moral *agents* without any explicit presumptions about the ontology of moral *patient*-hood.

I address the second issue by way of the first, and the first by way of the dispute between pluralism and monism in environmental ethics. Ecosystems, as well as the space environment, can be devoid of the kinds of things to which anthropocentric ethical theories grant moral consideration, e.g., humans and their experiences of pleasure and pain. And ecosystems, as well as the space environment, are not

J.S.J. Schwartz (✉)
Department of Philosophy, Wichita State University, 1845 North Fairmount, Wichita, KS 67260, USA
e-mail: james.schwartz@wichita.edu

J.S.J. Schwartz and T. Milligan (eds.), *The Ethics of Space Exploration*, Space and Society, DOI 10.1007/978-3-319-39827-3_7

especially well understood, ethically and scientifically. Thus environmental ethics and space ethics raise similar questions—whether ecosystems or the space environment contain anything of value, and if so, whether existing ethical views must be revised or replaced in order to accommodate these values. If it can be argued that the conventional preference for monism is contestable in the case of present-day environmental ethics—something I argue is the case in §2—then, *mutatis mutandis*, a similar case can be made that the conventional preference for monism is also contestable in the case of present-day space ethics.

The moral for space ethics I draw in §3 is that, conservatively, we ought to remain neutral with respect to the monism/pluralism dispute, at least until we have more robust knowledge about what is at stake in space, and about what kind of moral theory most cogently helps us to protect what is at stake. No amount of theorizing about values will eradicate the practical difficulties associated with making precise predictions about the shape of our duties in space environments—especially those that we know relatively little about. Less cautiously, there are practical *advantages* to pluralism. The methodology of pluralism encourages openness to diverse conceptions of value, and this coheres with the multi-faceted way in which the space environment is actually studied. Pluralism can also help us to be tolerant of reasons—reasons not *presently* appreciated by monist views—for conserving and preserving elements of the space environment. Thus, even if monism is an ideal we should strive for in ethics, its attainment in mature settings might well depend on a lengthy period of pluralist challenges at earlier stages.

In §4 address the second issue—of whether space ethics should even concern itself with intrinsic value—where I discuss Tony Milligan's recent development of an alternative type of position, emblematic of the attempt to conceive of moral agency without explicit presumptions about moral patients. I argue that Milligan's position faces practical difficulties equivalent to those faced by intrinsic value views—our relative ignorance about the space environment is also inimical to making substantive predictions about moral agency in space.

The upshot is not so much a defense of any particular position but rather an emphatic reminder that the challenges posed to ethics by the space environment will not be settled in a purely formal way and they will not be settled anytime soon. The most valuable contributions we can make at present come from offering proposals about how we might think about intrinsic value, virtue, etc., in space contexts. The more such proposals that survive the rigors of critical ethical discussion, the more prepared future generations of space ethicists will be, as more becomes known about living in space, for identifying and implementing fruitfully normative concepts to the space environment. This is, crucially, a process that should continue to evolve and grow in sophistication as our knowledge of the space environment evolves and grows in sophistication. The result is perhaps a process of theory construction that many normative ethicists might find unhygienic, but this is what space ethics, very much in its originary stages, demands.

7.2 Value Pluralism and Environmental Ethics

The dispute between monism and pluralism in environmental ethics is methodologically instructive for space ethics, largely because it helps to reveal the nature of the challenges that pluralism faces. But before I can explore these challenges I should first clarify my usage of the terms 'monism' and 'pluralism'. After all, one could espouse pluralism (and, by contrast, monism) about a variety of things, including: ethical theories; ethical principles; sources of ethical justification; intrinsic values; etc. A pluralist about *ethical theories* might hold that Kantianism is true in a restricted way, e.g., true for domains such as family and other interpersonal interactions, while in the same breath assert a restricted utilitarianism, e.g., for domains such as politics and economics. A monist, by contrast, would assert that at most one of these theories is true without restriction. A pluralist about *moral principles* might hold that the best moral theory incorporates an irreducible plurality of principles of right action—such as the categorical imperative (*sans* its Kantian superstructure), along with the greatest happiness principle (*sans* its Millian superstructure)—to be drawn upon when appropriate, whereas a monist would attempt to derive all duties from at most one of these principles. A pluralist about *ethical justification*, most commonly described as an ethical intuitionist, might hold that a plurality of ethical truths (the existence of an irreducible plurality of duties, for example) can be known directly via intuition. And finally, a *value pluralist* will hold that ethical conduct aims at the preservation or advancement of an irreducible plurality of basic or intrinsic values, whereas a monist aims to reduce all intrinsic value claims to a single basic value.[1]

My focus shall be on the dispute between monists and pluralists about *value*, and all uses of 'monism' and 'pluralism' below should be understood in this sense unless otherwise indicated. Although I will remain agnostic throughout about the specific metaphysical content of intrinsic value claims, nevertheless I am comfortable defining the term in a functional way, viz., that '*x* is intrinsically valuable' means "*x* is good in itself." The intended implication here is that something's being good in itself constitutes a *defeasible* but *independent* reason for preserving *x*, and perhaps also a reason for bringing *x* about. Thus I shall frame the dispute between monists and pluralists as a debate over whether there is just one intrinsic value, i.e., one single way something could be good in itself, or whether there is an irreducible plurality of intrinsic values, i.e., a plurality of ways things can be good in themselves.

---

[1]In none of these cases should pluralism be confused with *relativism*, the view that ethical claims are subjective. Each form of pluralism described here is offered as an objective ethical position.

7.2.1 Stone Against Monism

An early challenge to monism was posed by Christopher Stone, who argued that monist decision-making failed to adequately capture the diversity of concerns we employ in ethical deliberation. Although we are unlikely to ignore both utility and respect for persons, especially while deliberating about strictly interpersonal or strictly intersentient activities, nevertheless our deliberations often involve a much broader range of entities:

> It is only when one starts to wonder about exotic clients, such as future generations, the dead, embryos, animals, the spatially remote, tribes, trees, robots, mountains, and art works, that the assumptions which unify ordinary morals are called into question. (Stone 1988, 144)

Stone does not see a compelling reason why we should limit ourselves to just one moral theory, especially when we must then apply that theory to actions not just between rational beings or between sentients but also to actions which affect the dead, mountains, or ecosystems. Indeed, if the various entities Stone identifies differ not just in their degree of moral considerability but also in the ways they can be affected by our actions, then *prima facie* it is inappropriate to reduce the ethical evaluation of *all* action to a single principle drawn from a single underlying view of what is valuable. More sensible, thinks Stone, is to recognize that the diverse milieus in which we deliberate compel our appropriation of equally diverse values and principles. Thus, it may be that Kantianism is most appropriate for interpersonal relations (such as promise-keeping and family obligations); that utilitarianism is most appropriate for public policy (where decisions impact large numbers of sentients); and that a theory such as Aldo Leopold's Land Ethic is most appropriate for actions affecting ecosystems.

Note that Stone's pluralism is not necessary a *value* pluralism but instead is either a *theory* or *principle* pluralism. Nevertheless Stone's central point can be restated in a way that captures the motivation for value pluralism:

> It is only when one starts to wonder about exotic clients, such as future generations, the dead, embryos, animals, the spatially remote, tribes, trees, robots, mountains, and art works, that the assumptions which unify ordinary ascriptions of intrinsic value are called into question.

It strikes me as artificial and ad hoc to jump immediately to the conclusion that all of these listed entities either possess (or lack) intrinsic value for the same reasons. Rather, it seems to me to be more natural to claim that value can arise in distinct ways. For instance, Kant's position seems broadly correct in that the rational faculties of ordinary humans give rise to some part of their value. But the value of, say, a mountain, is more sensibly credited to its beauty, its history, or its integrity. Whether each conception of value is brought under one heading—intrinsic value—or separate—intrinsic *human*-value and intrinsic *mountain*-value—these values still, I suppose, provide reasons—defeasible and with varying degrees of strength—to respect and preserve the existence of their bearers.

7.2.2 Callicott Against Stone

J. Baird Callicott has been a vocal critic of Stone's pluralism. One of Callicott's central concerns is, he argues, metaphysical in nature. To endorse Stone's view, according to Callicott, is to endorse mutually incompatible views about what the world is like. One the one hand, under utilitarianism, "[t]he only 'things' to which good and evil can attach…are, naturally, positive and negative private *subjective* psychological states" (1990, 114; Callicott's emphasis). Meanwhile, Kantianism assumes "a vintage European Enlightenment philosophy of human nature in which Reason (with a capital *R*) constitutes the essence of 'man'" (ibid., 114–115). These views are incompatible in the sense that, according to utilitarianism, *only* psychological states are of fundamental moral value, whereas, according to Kantianism, *only* rational beings are of fundamental moral value.

This inconsistency can be overcome if we allow that the utilitarian and Kantian value claims can be extracted from their theoretical origins. There is no obvious barrier preventing us from claiming that, e.g., psychological states can be "good and evil" without asserting that *only* psychological states can be so—likewise for rational beings. All Callicott establishes is that the reasons *traditionally* or *historically* given for justifying the value of rational agents (via Kant's moral metaphysics) conflict with the reasons *traditionally* or *historically* given for justifying the value of certain psychological states (via utilitarian metaphysics). But to render pluralism inconsistent would involve showing much more than this—viz., that there is *no* pair of cogent arguments (one for the value of rational beings; the other for the value of psychological states) that contain metaphysically consistent premises.[2]

Callicott does go on to correctly press the practical criticism that pluralist ethical deliberation is likely to encounter serious complications (ibid., 112). After all, pluralists such as Stone do not provide effective procedures for resolving cases involving competing values or duties. But Callicott later admits that these complications are also likely to arise in monist deliberation (ibid., 120). Why then does Callicott persist in objecting to pluralism if it does not raise any uniquely serious difficulties—either in its metaphysics or in its use in ethical deliberation?

I believe that Callicott's concerns about pluralist metaphysics are best described as symptoms of what he takes to be an underlying defect in the pluralist's *methodology*. Specifically, he intimates that certain important theoretical virtues are lacking under pluralism. One such virtue is when a theory helps us to perceive the world in novel, and often more fruitful ways. On this point he argues that it is doubtful that the environmental turn in ethics (the recognition that ecosystems are moral patients) is merely the result of a sudden acquisition of "keener senses for

---

[2]This, of course, does not establish that any pluralist view is in fact metaphysically consistent—only that Callicott does not succeed in proving the contrary. A substantive *justification* for a pluralist view—as opposed to its mere *defense* against Callicott's objections—would require a positive argument for the joint consistency of the view's basic value claims.

empirical moral properties," rather, he suggests, our awareness comes from the development of environmental ethical theories which are instead "newly acquired cognitive lenses which have reorganized what our senses have perceived all along" (ibid., 120). In other words, it is the development of environmental ethical *theories* which is responsible for our altered (and presumably, improved) conceptions of what is valuable, rather than the other way around. I take Callicott's corresponding criticism to be that pluralism amounts to the mere opportunistic endorsement of value claims. Thus, pluralism encourages thieving over honest toiling; for to pursue pluralism is to give up on the kind of metaphysical investigation of the moral universe which, according to Callicott, is so essential to the progress of ethics.

7.2.3 A Reply to Callicott

Callicott's "methodological" criticism as I have described it involves a misunderstanding about the process of theory construction. We develop new theories (and revise old theories) because we find existing theories deficient for any number of reasons, but most often because our current best theory cannot explain various new observations we have encountered. If we want to press the analogy between scientific and ethical theorizing, then the ethical correlate of recalcitrant *scientific observations* would have to be recalcitrant *ethical intuitions* (including intuitions about what is valuable). Thus a key ingredient of ethical theory construction is the prior existence of compelling ethical intuitions—intuitions that existing theories cannot accommodate in satisfying ways. The development of environmental ethics is an example of this process: An increasing number of philosophers and environmentalists detected something of value in the natural world, or came to feel a duty to preserve or protect the natural world—values and duties they could neither express nor justify adequately in the pre-environmental ethical era. Accommodating these intuitions and feelings required hypostasizing novel loci of value and novel duties, with these new values and duties later on ensconced in increasingly systematic moral theories. In brief, new moral intuitions and perceptions were the initial drivers of environmental theories, and not the other way around. (Which is not to deny that these pre-theoretic moral intuitions and perceptions went unmodified by subsequent theory construction or to deny that existing theories shaped their development.)[3]

[3]Nor would I deny that context—social, cultural, and technological—plays a role in how pre-theoretic intuitions about the environment are manifested and invoked. It is certainly possible that one might value environmental preservation but not go to the effort to construct an environmental ethic because it would be impracticable pending cultural and technical advancements. So although intuitions about environmental values may not necessarily have been novel, nevertheless their *use* as a foundation for ethical theorizing was (which is all I require). Thanks to Kelly Smith for helpful discussion on this point.

If it were true that pluralism is a purely reactionary perspective, or one that merely scavenges value claims from established monist theories, then perhaps we ought to be concerned that pluralism would stifle ethical progress. But I suspect the opposite is more likely. If we are to proceed under the assumption that there must be only *one* fundamental account of intrinsic value, or that there must be only *one* fundamental principle guiding ethical conduct, then this kind of "cognitive lens" could very easily turn into a cognitive *blindfold* preventing us from seeing new and improved ethical worldviews. This has been and remains a serious problem for environmental ethics. As anyone who has taught environmental ethics or who has described environmental theories to their colleagues has no doubt experienced, views like Kantianism and utilitarianism have a very firm grip on students' and philosophers' ethical intuitions, and it can be incredibly difficult to convince them to keep their minds open to non-instrumental, non-anthropocentric value claims. A pluralist, I suspect, would have comparatively less difficulty entertaining novel value claims.

Callicott's exhortation for a unified moral metaphysics (ibid.) seems even less well-motivated when our attention turns to the much more reliable arbiter of environmental ontology, ecology. If ecology delivered a unified and coherent view of what ecosystems are (in the sense of treating ecosystems as all belonging to a common *natural kind*) then perhaps it would be prudent in environmental ethics to regard ecosystems as the sole bearers of intrinsic value. But what ecology instead presents us with are numerous and varied descriptions of ecosystems—thus an environmental ethic founded on just one kind of description would be open the charge of arbitrariness.

As Wim van der Steen explains, ecological theories, which are products of our limited attention and abilities, must make "trade-offs" between generality, realism, and precision. As an illustration,

> …in population genetics biologists model genetic changes over time in populations. Such changes are affected by many different factors. General, realistic, precise models covering many factors tend to be untractable. In view of some specific purpose (for example, the preservation of concrete ecosystems), it may be wise to make models or theories as realistic and precise as one can, because one needs good predictions. Such theories, however, are anything but general: for example, they may hold only for certain types of populations. For other purposes, simplified models may suffice. Biologists typically work out a variety of different simplifications to arrive at a qualitative picture of the whole. (van der Steen 1995, 216–217)

Van der Steen's point is that theory formation in ecology is idiosyncratic; the object of one's interests will inform how one's ecological theories are constructed, and to what degree these theories are general, realistic, and precise. Moreover, in many of the applied sciences, ecology included, models are often used in creative, non-descriptive ways, for instance to *suggest* plausible hypotheses rather than to *confirm* those hypotheses or to provide decisive predictions.[4] Thus we cannot

[4]For a useful survey of this kind of use of modeling in science, see Sterrett (forthcoming).

always safely assume that the goal of ecological modeling and experimentation is in the first place to describe the actual features of ecosystems, or even to elucidate what it is to be an ecosystem. The upshot is that ecology does not present us with

> ...a grand overarching theory. The search for such theories is misguided, especially in ecology, the discipline which concerns environmental ethics most. Ecology typically works with arrays of overlapping models and theories dealing with similar phenomena... Pluralism is rampant; monism is out of the question in ecology. (van der Steen 1995, 217)

According to van der Steen, then, there is not some privileged perspective that ecology provides. Rather, ecology provides an array of "cognitive lenses" from which to view the natural world. Thus to "unify" environmental ethics by fixating on one kind of description of what an ecosystem is (or could be) is to base one's environmental ethics on an arbitrarily restricted view of ecology. Our awareness of various macro-entities we might select as the bearers of value (and to which we might refer using 'ecosystem') depends entirely on "the scale of space and time of our investigations—which depends on the purposes of specific research" (ibid., 218).

Van der Steen's point is not unique to ecology in the sciences. Scientific theories are always driven by our interests and are constrained by the observations we are capable of making. Though certain disciplines may be characterized by the pursuit of unified theories—theoretical physics, for instance—a sundry of other sciences aim instead to develop a deep understanding of particulars—geology, for example. Accepting differing scientific methods is permissible—indeed, unavoidable—because the aims of the various scientific disciplines are so different. There is nothing inherently suspect or stifling about investigation that, rather than constructing overarching theories to deliver new insights, instead relies on low-level insights and interests to drive modeling and theorizing.

Connecting this point to ethics, we might say that our desire to do the right thing no more forces monism upon us than our desire for knowledge forces the methods of physics on our knowledge-seeking. This much is echoed by Thomas Nagel:

> To look for a single general theory of how to decide the right thing to do is like looking for a single theory of how to decide what to believe. Such progress as we have made in the systematic justification and criticism of beliefs has not come mostly from general principles of reasoning but from the understanding of particular areas, marked out by the different sciences, by history, by mathematics...it is as irrational to despair of systematic ethics because one cannot find a completely general account of what should be done as it would be to give up scientific research because there is no general method of arriving at true beliefs. (Nagel 1991, 135–136)

To draw out Nagel's analogy further, I would say that the progress of ethics has not come only or primarily from general theories but rather from pre-theoretic consideration of particular "anomalies" such as animals and ecosystems—the very kind of consideration pluralism encourages. Thus, if there is something defective about the methodology of pluralism, then it must be for reasons other than those suggested by Callicott.

If the above paragraphs read more as a *defense* of pluralism rather than a ringing endorsement, then that is by design. For it could be that once all of the relevant data are in that we will devise a coherent means for unifying what were previously thought of as distinct intrinsic values. Indeed, a not insignificant marker of theoretical progress is when phenomena thought previously to be unrelated can be accommodated by a single explanation or a single theory, for instance when it was realized that electricity and magnetism are manifestations of a single underlying force. So a monist view may be something worth striving for in the ideal case, if only because the intellectual task of attempting to unify values might yield serendipitous insights (though we should not presume that unification is inevitable).

My goal, of course, is not to chart the texture of an ideal moral theory but rather the more practical matter of addressing what sorts of ethical pursuits we ought to encourage, and, by contrast, what sorts of ethical pursuits we ought to discourage. If Callicott is right then it is clear that pluralism is methodologically suspect and that we ought to discourage pluralist theorizing. But he is wrong in the first instance precisely because, even in the maturing discipline of environmental ethics, we are highly cognizant that we possess *neither* all relevant moral perceptions and intuitions *nor* all relevant ecological knowledge. That is not to say that the construction of monist views must wait for that indefinite future moment when we have sufficient data—by all means let us see where monism can take us at present. The point is rather that we should not only be tolerant of, but actively encourage, disparate ways of thinking about environmental value. This means we should attempt to develop and understand various conceptions of value as though they were unique, entertaining unification only when we have an understanding of what a particular conception of value commits us to, what duties it entails (if any), etc. Thus I agree with Anthony Weston concerning ethical theorizing at the "originary stage":

> …the process of evolving values and practices at originary stages is seldom a smooth process of progressively filling in and instantiating earlier outlines…originary stages are the worst possible times at which to demand that we all speak with one voice. Once a set of values is culturally consolidated, it may well be possible, perhaps even necessary, to reduce them to some kind of consistency. But environmental values are unlikely to be in such a position for a very long time. (Weston 1992, 333)

In this respect what holds in the case of environmental ethics holds to an even greater degree in the case of the ethics of space. Compared to our knowledge of Earth's environments, much less is known about space, so it is even more practically important to keep an open ethical mind concerning value and conduct in space. One way of ensuring this at the level of normative ethics is to openly explore alternative possibilities for thinking about value and conduct in space. And here there will be even more preconceptions to guard against than in the case of environmental ethics, as even the strongest environmentalist might still have trouble recognizing anything of intrinsic value where there is no life, e.g., in most (if not all) of the non-terrestrial environments in space.

7.3 Lessons for the Methodology of Space Ethics

If the remarks of the previous section are in order, then Callicott's methodological criticism of pluralism misfires: Not only does the tolerance of plural intrinsic values fail to hinder the progress of ethics, pluralism's openness to novel loci of value is likely to be one of the *drivers* of ethical progress. Meanwhile, monism about environmental value seems arbitrary precisely because our underlying scientific understanding of ecosystems is so fractured and multi-faceted. Similar critical features can be observed in the cases of space ethics and space science.

The scientific study of the space environment is a plural undertaking well beyond the pluralism van der Steen observes in ecology. Not only is there a great diversity of phenomena to study in space (and hence a vast plurality of possible items of interest), there also are a large number of scientific disciplines interested in these phenomena, including: astrobiology; astrochemistry; astrometry; astrophysics; cosmology; and Earth and planetary science. There is not one agreed upon description of the space environment but instead a large number of partial and potentially overlapping descriptions that vary in scale based on the interests of those making the descriptions. And the scope of interest can vary dramatically in the space sciences: the entire universe; a galaxy; a solar system; a star, planet, or other body; a planetary or lunar surface; a formation on a planetary or lunar surface; a volume of space; a sample of material from a planet, moon, or other body; etc. Given the diversity of perspectives that space science provides, it would be premature to claim that nothing we discover in the space environment could possibly challenge our beliefs about what is valuable.

Consider the implications of our ignorance about the space environment for planetary protection policies. Presently, protection policies focus almost exclusively on the prevention and mitigation of forward contamination of sites of interest in the search for life in the solar system. Though the sentiments behind these policies are laudable—originally, an awareness that careless exploration could jeopardize important research, and more recently, a growing appreciation of the moral considerability of extraterrestrial life—nevertheless they are very much outgrowths of past and present scientific interests. These interests are, in turn, largely a function of our knowledge about the space environment. But as this knowledge expands, there will likely be dramatic changes in the class of phenomena that excite our scientific curiosity. It is doubtful that our moral sensibilities and intuitions will go unmodified during this expansion of our knowledge. The protection of extraterrestrial life may always be one of the strongest ethical duties for space explorers, but if too much emphasis is placed on life we may blind ourselves to reasons for preserving other kinds of environments. As evidenced by the impact on environmentalism following *Mariner 2*'s confirmation of Venus' runaway greenhouse effect, we cannot always predict our future scientific interests, and, in turn, the ethical consequences of those interests. To preserve such opportunities for scientific discovery we would be better served by ethical views which are *more* rather than *less* amenable to novel reasons for environmental protection.

Let me be clear that I am not suggesting that the methods of ethical theorizing should be replaced with more characteristically scientific methods or that the ethics of the space environment will somehow be reducible to its scientific examination. Nor am I claiming that space science has any special authority over ascriptions of value in the space environment. What space science does have authority over is the ontology of the space environment, i.e., the class of entities that we *select from* when deliberating ethically about what is intrinsically valuable (just as ecological science, *inter alia*, has authority over the ontology drawn upon in environmental ethics). Thus I am only making the naturalistic, metaphysical claim that any theory of value needs to be responsive to *real* entities and *real* features of the space environment. This is ethically salient information because the class of entities we take to fall under the scope of moral consideration, or to have intrinsic value, is very much a function of what we know about what exists and about what we know about the relationships that obtain between those entities. Moreover, if Holmes Rolston is correct, then the reverse holds as well—that the direction of our scientific investigations is, at least in part, a function of what we value:

> …the connection of description with evaluation is more complex, for description and evaluation to some extent arise together and it is often difficult to say which is prior and which is subordinate. Ecological description finds unity, harmony, interdependence, creativity, life support, conflict and complement in dialectic, stability, richness, community—and these are valuationally endorsed, yet they are found, to some extent, because we search with a disposition to value these things…ecological description does not merely confirm these values; it informs them, and we find…the empirical content, of order, harmony, stability, richness, community is drawn from no less than brought to nature. (Rolston 1988, 231)

Whether Rolston's list of descriptions/values is empirically correct can and perhaps should be challenged, nevertheless we should not ignore his basic insight—that what we find worth studying and what we value are highly interdependent.[5]

To summarize, our ethical sensibilities should be informed by our scientific understanding of the world. It would be a tremendous mistake to assume that our understanding of the space environment has progressed to the point that further exploration will never modify our ethical thinking in a fundamental metaphysical way, by, e.g., identifying a new class of entities or a new way of looking at a system. Methodologically, the ethics (and science) of space has much to benefit from the toleration of pluralism about value. As a corollary, pluralism in ethics encourages a broader sympathy for space science, for pluralists should not only be interested in the findings of, e.g., astrobiologists (who might be the sole focus of biocentric monists), but also planetary scientists, astrophysicists, astronomers, etc.

---

[5]To be sure, the connection between scientific examination and valuation extends also to instrumental valuing, as our scientific knowledge also affects, and is affected by, what things we find useful for other purposes. But that does not negate such a connection between science and *intrinsic* valuing, which is Rolston's focus. Moreover, Rolston's perspective retains merit even if the connections are not necessary. What matters is that scientific study and valuation *can* (and often *do*) affect one another—not that they must do so in every single instance.

It is worth noting that the kind of pluralist view defended here is still open to Callicott's more practical criticism that pluralism unduly complicates ethical deliberation. Matters are not helped given the uncertainties surrounding our knowledge of the space environment and how this knowledge impacts our ethical evaluation of space. But since these same uncertainties are likely to hinder monist views as well, then perhaps we should recall that not all ethical frameworks are value-oriented—perhaps there is merit to the idea that our focus should be elsewhere, e.g., on moral *agents*.

7.4 What About Agent-Based Ethics?

Can we establish a more decisive deliberative framework for the ethics of space by shifting to a normative position that derives duties from our status as moral agents rather than various entities' status as moral patients? It is reasonable to have such a hope. After all, though we do not have a tremendous amount of experience in the space environment we do nonetheless have extensive knowledge of ourselves (even if that knowledge is imperfect). This, Saara Reiman claims, is where our focus should be:

> Estimating how one's intended actions realize virtues…can be done more reliably even in unexpected situations because conducting such an estimation does not require detailed knowledge about one's environment. The properties of the moral agent, not the properties of his/her environment are the key to doing the right thing, and no matter how strange the circumstances an explorer may find himself in, she always has access to knowledge about her own state of mind. Further knowledge can be useful, but it is not crucial to moral decision making in the way it is in consequentialist approaches. An explorer who has a grasp of virtue ethics may find that s/he should learn more about the situation s/he is in, but the capability to make moral decisions is never crippled to a level where making a moral choice is little more than a guess. (Reiman 2009, 86)

Might the development of space ethics be as straightforward as extending already-existing views, such as virtue ethics positions, which deal primarily in the examination of those kinds of character traits which are conducive to human flourishing? To see why pessimism is warranted here it is worth briefly examining one such position recently articulated by Tony Milligan.

Though Milligan does not describe his position as an instance of virtue ethics, he nevertheless has the similar aim of articulating a set of character traits or ethical dispositions that are essential to our humanity. His basic premise is that it is a feature of our humanity—of what it is to be human—that we belong to and act "as part of a community of values" (2015, 38). In other words, the mere fact of our humanity entails the existence of a variety of duties and obligations related to the maintenance of our humanity. (A virtue ethicist might by contrast claim that being aware of and acting on these duties is constitutive of certain human virtues.) One might object at this stage that our status as human beings is a purely biological fact of our existence and that this fact does not in itself entail any moral obligations, but this would, I think, miss Milligan's point. Accidental or not, it *is* a feature of our existence that we belong to a moral community and it is the case that this community shares certain values.

According to Milligan, one of our obligations is that "as we move into space we must preserve the sense of care and belonging which is integral to our humanity" (Milligan 2013, 4). What does it mean to preserve a sense of care and belonging? He elaborates while discussing the case of lunar mining:

> By all means we should still use the Moon, just as we use the Earth which we ourselves are part of, but if we *merely* use it, or regard it *merely* as a resource then we risk losing sight of the ethical task of *being at home* in a larger region of space, a task which is different from learning to live in or near to a mining district. The places where we dwell, if we are truly to *dwell* anywhere and not merely to *exist*, must have meaning and significance for us. (ibid.)

What seems to be important here is having a sense of care and belonging to the places we reside, i.e., our *homes*—and it does not matter whether these are terrestrial homes or extraterrestrial homes. As the example illustrates, part of what having a sense of care and belonging involves is viewing one's home as *something more* than a mere resource, and as a place where one is *not* free to be destructive and wasteful (ibid., 5).

Aside from his prediction that strip-mining the moon for He_3 would have to occur on such a scale as to unveil our attitudes as wholly uncaring and alienating, Milligan is cautious in drawing out any wide-ranging, definitive practical consequences. Rather, he is sensitive to our paucity of experience with living in space:

> To do justice to what it is to live a life like ours we need to recognize that there is a wide range of ethically-sensitive concepts which play a role in our best descriptions, often without drawing attention to themselves: betrayal, fairness, cruelty, humility, courage and humanity itself. The list may not be a short one. (Milligan 2014, 47)

Moreover, he acknowledges that "the list" of relevant moral considerations will likely undergo modification in future to better suit ethical deliberation in space:

> It is also unlikely to contain all of the conceptual resources that we will need for radically new situations such as those humans will face off-world, even in those nearby regions of space which we will move into in the coming decades. There may always be room for another concept or a modified concept, one which helps us to understand what it is to be human, to be bound to other humans and to other creatures, or which helps us to understand what it is to be connected in deep ways with some place or other: a village, a country or a home-planet. (ibid., 47–48)

Thus in at least one respect Milligan's position does not make clear gains in predicting the shape of the ethics of space: His position, like the value-oriented strategies discussed in previous sections, relies critically on how the knowledge we gain living in and exploring space will impact what we view as salient moral considerations. But this much holds in the case of a scientifically-informed pluralism as well.

Nevertheless Milligan's position contains a number of familiar elements and could be advantageous for this reason. Certainly our understanding of betrayal, fairness, cruelty, etc., will not undergo unrecognizable modifications, and our present understanding of these concepts will be easily imported to actions in space.

I agree up to a point. But even familiar notions and concepts are likely to manifest themselves in unexpected ways in the space environment. For instance, our intuitions about what it is to call a place home, what we take to be destructive and wasteful behavior, etc., are very much wedded to our experiences on the home planet. And, as Milligan admits, we have good reasons for supposing that extraterrestrial cultures will be profoundly different than most, if not all, terrestrial cultures. Just what wasteful and destructive behavior will look like in space will no doubt be radically different from what these activities look like on Earth (where even here there is a dramatic variety in what cultures consider to be destructive and wasteful).

I do not mean to suggest that Milligan's position is incoherent or involves internal inconsistencies. Rather, I wish only to identify that it faces its own versions of the practical problems that value-oriented views face: Our evident lack of knowledge about the space environment, coupled with our lack of experience living in the space environment, makes it practically impossible to predict the precise shape of the duties involved in Milligan's position.[6] That is not to say that dramatic improvements in our knowledge of the space environment will automatically resolve the present difficulty, nor that they will identify a unique best ethical theory for space. It is only to emphasize, as Milligan is keenly aware, that shifting focus away from moral patients back on to agents is not some kind of ethical magical wand that allows us to make coherent, rational ethical decisions in situations of great uncertainty. Contrary to Reiman, sometimes we must learn more before we have a right to describe our decisions as 'ethical'. Of course, value-oriented ethicists can make a similar retort—sometimes we must learn more before we have a right to say what is and is not intrinsically valuable. There is no compelling reason at present to pursue one strategy at the exclusion of the other.

7.5 Conclusion

In space ethics the tricky questions are *pervasive*. Not only does our lack of knowledge about the space environment impede us from speaking with confidence about intrinsic values in space, it also hinders us from speaking with confidence about virtue and duty in space. This gives rise to two inveterate features of the methodology of space ethics: that it must develop slowly and alongside our pursuits in space, and that it should function as a reflective guide for these pursuits. Not only will our actions in space guide how we think about space ethics—space ethics will guide decisions about what we do in space.

Acknowledgments Thanks to Mukesh Bhatt, Noell Birondo, Tony Milligan, Christopher Newman, and Kelly Smith for discussion and comments.

[6] I would also argue that other agent-based positions available face their own versions of the "lack of knowledge" problem, e.g., McArthur and Boran (2004), Sparrow (1999), and Sparrow (2015).

References

Callicott, J. B. (1990). The case against Moral Pluralism. *Environmental Ethics, 12*, 99–124.
McArthur, D., & Boran, I. (2004). Agent-centered restrictions and the ethics of space exploration. *Journal of Social Philosophy, 35*, 148–163.
Milligan, T. (2013). *Scratching the surface: The ethics of helium-3 extraction*. Paper presented at the 8th IAA symposium on the future of space exploration: Toward the stars, Torino, Italy.
Milligan, T. (2014). *Nobody owns the moon: The ethics of space exploitation*. Jefferson, NC: McFarland & Company Inc, Publishers.
Milligan, T. (2015). Fear of freedom: the legacy of Arendt and Ballard's space skepticism. In C. Cockell (Ed.), *The meaning of liberty beyond earth* (pp. 33–45). Springer International.
Nagel, T. (1991). The fragmentation of value. In T. Nagel (Ed.), *Mortal questions* (pp. 128–141). Cambridge: Cambridge University Press.
Reiman, S. (2009). Is space an environment? *Space policy, 25*, 81–87.
Rolston, H. (1988). *Environmental Ethics: Duties to and Values in the Natural World*. Temple University Press.
Sparrow, R. (1999). The ethics of terraforming. *Environmental Ethics, 21*, 227–245.
Sparrow, R. (2015). Terraforming, vandalism, and virtue ethics. In J. Galliott (Ed.), *Commercial space exploration: Ethics, policy and governance* (pp. 161–178).
Sterrett, S. (forthcoming). Experimentation on Analogue Models. In L. Magnani & T. Bertolotti (Eds.), *Springer handbook of model-based science*. Berlin: Springer.
Stone, C. D. (1988). Moral pluralism and the course of environmental ethics. *Environmental Ethics, 10*, 139–154.
van der Steen, W. (1995). The demise of monism and pluralism in environmental ethics. *Environmental Ethics, 17*, 209–220.
Weston, A. (1992). Before environmental ethics. *Environmental Ethics, 14*, 321–338.

Chapter 8
The Ethics of Outer Space: A Consequentialist Perspective

Seth D. Baum

8.1 Introduction

Consequentialism maintains that individuals should act so as to achieve good consequences. It is typically, but not necessarily, structured as an exercise in optimization: individuals should act so as to achieve the best consequences that they are able to bring about. Consequentialism is one of the most prominent and widely supported forms of ethics, though it is also often criticized (e.g., Scheffler 1982, 1988; Glover 1990). In this paper I will discuss the significance of outer space for a range of consequentialist views.

Most treatments of consequentialism focus on consequences located on Earth. When people ask, "What are the consequences of my actions?", they usually do not consider the consequences for other planets or faraway stars or galaxies. Research using consequentialist frameworks has a similar terrestrial focus. This holds even for big-picture consequentialist research like cost-benefit analyses of climate change actions, which consider consequences for the entirety of Earth and many years into the future but nothing beyond Earth (e.g., Stern 2006; Nordhaus 2007). This terrestrial focus may seem reasonable, but it is a mistake.

Outer space is important for consequentialism for at least two reasons. First, the vast expanses of outer space offer opportunities for achieving vastly more good or bad consequences than can be achieved on Earth alone. Actions that bring good consequences to some significant portion of the universe are of exceptionally high value in many consequentialist frameworks.

The second reason why outer space is important for consequentialism is the possibility that humanity may encounter an intelligent extraterrestrial civilization

S.D. Baum (✉)
Global Catastrophic Risk Institute, Seattle, USA
e-mail: seth@gcrinstitute.org

J.S.J. Schwartz and T. Milligan (eds.), *The Ethics of Space Exploration*, Space and Society, DOI 10.1007/978-3-319-39827-3_8

(abbreviated ETI for extraterrestrial intelligence). Conspiracy theories aside, no such civilization has yet been detected. It is possible that none exist. However, the search for extraterrestrial intelligence (SETI) remains a young endeavor. As it proceeds, and as humanity progresses and expands as a civilization, the odds of an encounter increase. ETI encounter poses certain challenges for consequentialism and can also yield some dramatic consequences.

This paper contributes to a modest body of prior literature on outer space consequentialism. For example, the vast opportunity for human colonization of outer space to bring is discussed in Ng (1991), Tonn (1999, 2002), Ćirković (2002), Bostrom (2003), and Baum (2009, 2010a). Additionally, Rolston (1986), Fogg (2000), Cockell (2005), Haqq-Misra (2012), and Milligan (2015) consider whether space colonization would be a good consequence. Finally, Cockell (2007), Baum (2010b), and Baum et al. (2011) discuss the ethics of ETI encounter. These publications do not all endorse consequentialism or focus exclusively on consequentialism but nonetheless contain relevant discussion.

8.2 Consequentialism Specifics

There are many different consequentialist ethical frameworks, with varying implications for outer space. This section presents some different consequentialist frameworks in terms of what they intrinsically value and how they weight intrinsic value across species and spatiotemporal location. Sections 8.3 and 8.4 then discuss implications of the different consequentialist frameworks for outer space. This section also briefly argues for my preferred form of consequentialism. My arguments are likely too brief to be widely convincing and are intended simply to let the reader know where I stand.

Before diving into the specifics, here is a brief defense of consequentialism. This is also not intended to be widely convincing, only to hint at consequentialism's appeal. The appeal is that consequentialism recommends whatever is best for the world or the universe, however "best" is defined (more on that below). Some critics argue that certain actions are required or forbidden regardless of the consequences. But suppose, for example, that torturing someone would somehow spare many other people from torture. However deplorable torture may be, it should not be categorically forbidden, nor should anything else. Other critics argue that at least some circumstances cannot be resolved by evaluation of consequences, and instead must be resolved through use of "practical wisdom" or virtue. But then one could claim, for example, that refusing to torture someone is virtuous, again even if it results in many other people being tortured. Appeals to virtue do not excuse causing worse outcomes. In short, consequentialism succeeds by considering all of the possible outcomes of all of the possible options and recommending whichever option is best.

8.2.1 Intrinsic Value

Something is intrinsically valuable if it is valuable for its own sake (Rønnow-Rasmussen and Zimmerman 2005). In contrast, something is extrinsically valuable if of its value derives from its relation to intrinsic value (Bradley 1998). The most commonly discussed type of extrinsic value is instrumental value; something is instrumentally valuable when it is means for, or a cause of, intrinsic value. For example, if human health holds intrinsic value, then medicine can hold instrumental value.

Consequentialism ultimately seeks to increase intrinsic value. A consequentialist may pay attention to extrinsic value, but this is only because it affects intrinsic value. Intrinsic value is so central to consequentialism that specific frameworks are often known in terms of what they consider to hold intrinsic value. Here are some important examples:

- *Subjective experience* is the cognitive sensation had by sentient beings, essentially how something is perceived to feel. A consequentialist framework that intrinsically values subjective experience essentially says "if it feels good, it is good". This is one form of utilitarianism (e.g., Tännsjö 1998; Ng 2003). Other terms used for subjective experience include welfare, wellbeing, and quality of life. The feeling of subjective experience need not be a crude, instantaneous pleasure; instead, it can be a more complex overall life experience. Indeed, for some people, periods of adversity are found to bring overall life benefits (Seligman 2010).
- *Preference satisfaction* occurs when individuals get what they want. It is distinct from, but often conflated with, subjective experience. The two differ in that individuals do not necessarily prefer improvements in their own subjective experience. Preference satisfaction holds intrinsic value in the other major form of utilitarianism (e.g., Barry 1989; Adler and Posner 2000). It also underlies some consequentialist forms of procedural justice, such as voting schemes for making social choices, on the premise that people vote for what they prefer.
- *Ecosystem flourishing* occurs when ecosystems are healthy and productive and biologically successful. A single biological organism would have a small amount of ecosystem flourishing on its own; an intricate community of organisms would have more. Ecosystem flourishing is used often in wildlife management (Niemi and McDonald 2004), and some argue that it should be the basis for all ethics (Holbrook 1997). Consequentialist ecocentrism places intrinsic value on ecosystem flourishing and argues that individuals should act to increase it.
- *Abiotic systems* contain no life. In this context, abiotic systems are usually understood to also contain no sentience or intelligence, meaning that they exclude abiotic artificial intelligence (AI) . The intrinsic value of abiotic systems is especially important for extraterrestrial settings because they are not known to contain any life. An important form of abiotic consequentialism holds that intrinsic value is diminished when life or AI intrudes on an undisturbed abiotic

system (Rolston 1986; note that Rolston does not discuss AI but the same logic applies to both AI and life).

A particular consequentialist framework can place intrinsic value on several or all of these phenomena, and/or other phenomena as well (Baum 2012). For example, Milligan (2015) proposes that abiotic or microbial extraterrestrial environments should be colonized by humans only if humans had sufficient reason to do so. While Milligan (2015) avoids purely consequentialist reasoning, the logic nonetheless parallels what a consequentialist framework might recommend if it intrinsically values humans as well as abiotic or microbial extraterrestrial environments.

I would argue for placing intrinsic value exclusively on subjective experience. The simple reason is that only subjective experience feels good. The good feelings we experience give us reason to be glad we exist, and likewise give us reason to be glad for the existence of other beings that have positive subjective experiences. Ecosystems and abiotic systems might seem nice to us, but they cannot enjoy their own existence. Their existence brings them no joy or happiness or any other subjective experience, without which they have (I would argue) no reason to be intrinsically valued. The argument against preference satisfaction is subtler and not crucial for this paper.

8.2.2 Species

Should intrinsic value be weighted equally across all species? For example, is the subjective experience of a chicken worth as much as that of a human? Most treatments of consequentialism only place intrinsic value on humans, i.e. their ethics are *anthropocentric*. Some count all intrinsic value equally regardless of species; Baum (2010b) calls this *universalist*, but in this paper I will use the term *egalitarian* because its core feature is the equal weighting of intrinsic value. A fully egalitarian framework must also weight equally across location (Sect. 8.2.3). I will use the terms *species-egalitarian* and *location-egalitarian* for frameworks that are egalitarian with respect to species and location, and the unqualified term *egalitarian* for frameworks that are egalitarian with respect to both.

I would argue for species-egalitarianism. It is an arbitrary coincidence which species anyone happened to be born into. The fact that we happen to be humans is likewise an arbitrary reason to favor humans. If we favor humans, it is probably because we are biased. The pleasure of another species is still pleasure, and should count as much as that of a human, adjusted for its intensity and duration. The same point holds for preference satisfaction.

Species-egalitarian frameworks can still place more intrinsic value on members of certain species. A typical human life might have more positive subjective experience or preference satisfaction than a typical chicken life, simply because humans live longer and are more cognitively advanced. A species-egalitarian utilitarian framework would thus typically favor saving the life of a human over the life of a chicken, even while it would oppose acts such as causing a chicken to suffer immensely just so a human can eat chicken or eggs instead of a perfectly good vegan meal.

8.2.3 Location

Should intrinsic value be weighted equally across all locations in space and time? For example, is the life of a human worth more if it occurs here-and-now instead of there-and-later? Ethical frameworks that weight intrinsic value differently across locations can be said to *discount* intrinsic value (Price 1993; Portney and Weyant 1999; Perrings and Hannon 2001). A location-egalitarian consequentialist framework would weight all intrinsic value equally regardless of where or when they occur.

It is sometimes argued that intrinsic value should be discounted across space and time, for example because people have special duties to those closer to them (Cowen and Parfit 1992; Smith 1998). However, I would argue for location-egalitarianism. The argument here has the same logic as my argument for species-egalitarianism: it is an arbitrary coincidence which location we happened to be born into, and whatever phenomenon is intrinsically valued is still the same phenomenon regardless of its location.

Location-egalitarian frameworks can still favor some locations over others for instrumental reasons. We are often more capable of helping those near us than those distant from us. For example, I might help carry my neighbor's groceries, but I will not do the same for someone in another city. This does not mean I place more intrinsic value on my neighbor. It just means that, given my own instrumental capabilities, I can bring about more intrinsic value by helping my neighbor.

8.3 Opportunities in Outer Space

Earth is limited in both space and time, whereas the rest of the universe is much larger. Intuitively, this would suggest large amounts of intrinsic value could be accrued in outer space, especially under a location-egalitarian framework. However, this does not necessarily mean that outer space holds great opportunities for humans.

The consequentialist argument for space colonization has two parts. First, it must be the case that space colonization brings an increase in intrinsic value. This is to say that space colonies would be an improvement over whatever would otherwise exist. A colony on Mars, for example, would be better than an uncolonized Mars. Second, the improvement must be large enough to justify the effort it takes to colonize space. In other words, space colonization must not be so difficult that there are better options to increase intrinsic value on Earth. A Mars colony, for example, could be quite expensive. That same money could be used for other purposes, for example to reduce poverty. Only if space colonization is the best use of available resources can a consequentialist case for it be made. This section evaluates these two parts for the range of consequentialist frameworks.

8.3.1 Does Space Colonization Increase Intrinsic Value?

Whether space colonization is an improvement depends firstly on what holds intrinsic value. Space colonization changes the character of extraterrestrial environments from (apparently) abiotic systems to places populated with intelligent life. Space colonization is thus more likely to be an improvement if intrinsic value is placed on some aspect of intelligent life, such as subjective experience or preference satisfaction.

Space colonization would also typically be an improvement if ecosystem flourishing holds intrinsic value. Even a single human body contains a diverse ecosystem of microbes plus the human life. A single human could thus bring a significant increase in ecosystem flourishing to an uninhabited planet, depending on the precise formulation of "ecosystem flourishing." For an inhabited planet, the net effect depends on how the colony affects the indigenous biota. Cockell (2005) explains that if the indigenous biota is intrinsically valued, then space colonization can still proceed if the colony does not harm the biota; furthermore, space colonization should proceed if the colony can help the biota, such as by bringing it nutrients that can help it live and thrive. This is a reasonable position, but it does not resolve the issue of how to make tradeoffs between indigenous and colonizing ecosystems. Is it ever permissible for space colonization to proceed if it harms indigenous biota? The ecocentric consequentialist would say yes, it is permissible if it results in a net increase in ecosystem flourishing.

If intrinsic value is placed exclusively on abiotic systems, then space colonization is unlikely to increase intrinsic value. Indeed, it may even decrease intrinsic value, especially if undisturbed abiotic systems hold greater intrinsic value. This logic is found, for example, in the preservation ethic of Rolston (1986, 170), arguing against space colonization on grounds that "humans ought to preserve projects of formed integrity, wherever found". Fogg (2000) makes a compelling counterargument on grounds that humans are part of nature, meaning that there is no clear baseline state of nature that should be preserved.

An instrumental argument can be made against some space colonization, even if the extraterrestrial body holds no intrinsic value. Cockell (2005) describes that (1) an undisturbed extraterrestrial body can be instrumentally valuable, for example as an object of scientific study, and (2) visiting the body can be harmful to humans or to Earth, for example if it infects Earthlings with an extraterrestrial disease (i.e., back contamination). Taking these instrumental factors into account, space colonization should not be completely avoided, but it should proceed with caution. Colonization permanently ends billions of years of isolation. Humanity should seek to realize what instrumental value exists in undisturbed extraterrestrial environments before it is too late.

What about different weightings of species? An anthropocentric ethics would tip the scales even further in favor of space colonization. In anthropocentric ethics, space colonization would be an improvement unless it harmed humans, for example via back contamination. The same logic applies to an ethics that only intrinsically values Earth species. Alternatively, an ethics that only places intrinsic value on extraterrestrial species would be categorically against space colonization if there is even any chance that extraterrestrial species could be harmed. Meanwhile, a species-egalitarian ethics would tend to favor space colonization, unless it only intrinsically values abiotic systems.

Finally, on different weightings of location: An ethics that favors nearby locations in space and time would diminish the intrinsic value of space colonization. But as long as distant locations still hold some intrinsic value, space colonization could still be an improvement. However, if an ethics only intrinsically values locations on Earth, then the only potential value of space colonization would be instrumental. For example, an asteroid mining industry would be free to pollute outer space as long as it improves conditions on Earth. This would suggest little merit for space colonization. An Earth-only ethics also implies that the lives of astronauts hold no intrinsic value while they are in outer space. As with species-egalitarianism, a location-egalitarian ethics would tend to favor space colonization.

In summary, consequentialist ethics will typically conclude that space colonization is an improvement. The core exceptions will be specific cases in which the colony harms the colonists or in which indigenous biota are harmed more than the colonists are helped.

8.3.2 Is Space Colonization Worth the Effort?

Space colonization may bring an increase in intrinsic value, but it is of course not the only activity that can. Consequentialism typically argues that individuals should act so as to bring about the best consequences, meaning the largest increase in intrinsic value. Consequentialism will thus only recommend space colonization if it brings the largest increase *per unit effort* relative to other options. Otherwise, our effort is better spent elsewhere.

Space colonization is notable because it may be able to bring utterly immense increases in intrinsic value. Early colonies might start small, given that other planets and moons have inhospitable environments. However, it may be possible to build large indoor colonies or create more hospitable outdoor environments (i.e., terraforming). Even just on other planets and moons in the Solar System, space colonies could multiply the total area available for human habitation. And there are many more planets around other stars, as ongoing research on exoplanets is now learning. One recent study estimates 22 % of Sun-like stars have Earth-like exoplanets (Petigura et al. 2013), implying billions to tens of billions of potentially habitable planets across the galaxy.

Opportunities at any given star may also be quite a bit greater than those available only on planets. Earth only receives about one two-billionth of the Sun's radiation. To collect all the Sun's radiation, humanity would need a Dyson swarm (named after Dyson 1960), which is a series of structures that surrounds a star, collecting its radiation to power a civilization. A Dyson swarm around the Sun could potentially enable a civilization a billion times larger than is possible on Earth. Likewise, Dyson swarms around one billion stars would bring humanity approximately 10^{18} (one billion–billion) times more energy *per unit time*.

Space colonies could also increase the amount of time available for human civilization. Earth will remain habitable for a few billion more years (O'Malley-James et al. 2014). Stars will continue shining for about 10^{14} more years (Adams 2008). That gives us an additional 10^5 times more energy, for a total of 10^{23} times more energy than is available on Earth. After the stars fade, other energy sources may be available. And even if our current universe eventually becomes uninhabitable, it may be possible to move to other universes (Kaku 2005). The physics here is speculative, but it cannot be ruled out, and hence there is a nonzero chance of a literally infinite opportunity for space colonization (Baum 2010a).

Whether the opportunity is infinite or merely, say, 10^{23} times larger than what can be done on Earth, the opportunity is clearly immense. As long as space colonization is an improvement (Sect. 8.3.1), then it would seem that the consequentialist should prioritize space colonization. The sooner space colonization begins, the more of its immense opportunity can be gained. Indeed, Ćirković (2002) estimates 5×10^{46} human lifetimes are lost for every century in which space colonization is delayed.

There can also be large value for space colonization under ecocentric intrinsic value. It is sometimes argued that Earth would be better off without humans. For example, the Voluntary Human Extinction Movement states that "Phasing out the human race by voluntarily ceasing to breed will allow Earth's biosphere to return to good health" (http://vhemt.org, accessed 25 October 2015). However, this makes sense only if extraterrestrial locations are not intrinsically valued. Otherwise, exterminating humanity ruins the opportunity for humans to bring flourishing ecosystems into outer space. Terraforming other planets or bringing ecosystems into Dyson swarms could bring immense amounts of ecosystem flourishing.

What about the cost of space colonization? It is true that space missions are very expensive. Today's space missions commonly cost billions of dollars, and these are small relative to what would be needed for a Dyson swarm or galactic civilization. A cosmic civilization might bring 10^{23} times more opportunity, but if it also costs 10^{23} times more than, say, reducing poverty, then it might not actually be worthwhile.

The exact cost of building an immense cosmic civilization cannot be calculated given present knowledge. Indeed, it is not presently known whether building it is even possible. Building it would involve technologies and procedures that do not yet exist and cannot yet be costed. Therefore, the cost-effectiveness of space colonization cannot yet be estimated with any precision. (This holds whether cost is measured in terms of money, effort, or anything else.) This ambiguity would seem to be stifling for consequentialism.

The ambiguity can be resolved by the fact that space colonization is a long-term project. Humanity is just not going to build an immense cosmic civilization any time soon. Today's most ambitious plans call for tiny extraterrestrial colonies such as the Mars One project for a permanent settlement on Mars. Such colonies are a far cry from the immensity of a galactic civilization. Building and populating something immense will take a lot of time.

It therefore follows that actions today should focus not on immediate colonization per se, but instead on enabling colonization sometime in the future. This can be done, for example, by developing technologies that can be used for space colonization. Such technologies lower the cost of colonization so that it eventually becomes worth the effort. Transformative future technologies such as atomically precise manufacturing (Drexler 2013) are especially worth pursuing.

Another productive means of enabling future space colonization is by keeping human civilization intact. Space colonization cannot proceed if human civilization does not exist. Even if humanity is not completely extinct, it will take an advanced civilization to colonize space. Threats such as nuclear war, pandemics, global warming, asteroid impacts, and supervolcano eruptions are among the threats that have the potential to knock human civilization out. Thus, if the goal is eventual space colonization, an essential priority is avoiding civilization's collapse (Asimov 1979; Ng 1991; Tonn 1999, 2002; Bostrom 2003; Baum 2009, 2010a).

For several reasons, the main priority today should be keeping human civilization intact. First, civilization today faces an alarmingly long list of alarmingly urgent threats. Our survival is hardly guaranteed. Second, any other activities we might pursue depend on civilization's continued existence. Advanced technology, space exploration, colonization—all of these things can be pursued at any time as long as civilization still exists. While there is large value in hastening space colonization (Ćirković 2002), the value of avoiding destruction may be even larger. Whether it is larger depends on the details, which could take a lot of work to evaluate. A third reason is thus to continue evaluating the best options and charting its course. Thus, while space colonization may be worth pursuing now under certain exceptional circumstances, under most circumstances the priority should be keeping civilization intact.

8.4 Extraterrestrial Intelligence Encounter

At the end of Sect. 8.3.1, I concluded that consequentialism will typically find space colonization to be good unless it harms the colonists or if it harms indigenous biota more than colonists are helped. The possibility of encounter with indigenous biota—with extraterrestrial life—could result in harm to either. It could also result in benefit to either. But which would it be? And what does that mean in practical terms for human actions right now? This section answers these questions for the range of consequentialist frameworks and specifically for encounter with extraterrestrial intelligence, ETI. Non-intelligent extraterrestrial life is also important, but ETI have especially meaningful implications in the context of many forms of consequentialism.

8.4.1 Encounter Scenarios

Encounter with extraterrestrials is a common theme in science fiction, but its nonfiction study is fraught with uncertainty. We humans just do not know what extraterrestrials would be like and how they would react to us. We thus should resist the temptation to predict how an encounter would proceed. Instead, the best we can do is to consider a range of scenarios and logically evaluate the consequences of each.

Surveys of encounter scenarios are provided in Michaud (2007) and Baum et al. (2011). Baum et al. (2011) use an anthropocentric consequentialist perspective, assessing whether encounter would benefit or harm humanity. A core point is that an ETI is likely to be more advanced than humanity, because human civilization is young relative to astronomical time scales. That means that the ETI would likely (but not necessarily) get what they want. For this reason, it is important to consider what they might want—that is, what their ethics are.

Competing narratives of extraterrestrial encounter assume that ETI would either seek to help or harm humanity. The "seek to help" narrative tends to emphasize humanity's more altruistic or egalitarian tendencies. It often posits that more advanced civilizations will tend to be more egalitarian, or at least that non-egalitarian civilizations are more likely to self-destruct from misuse of dangerous technologies (Sagan and Newman 1983). The "seek to harm" narrative tends to emphasize humanity's more selfish and anthropocentric tendencies. It often observes that colonization throughout history frequently ends poorly for the colonized (Diamond 1999).

In general, one might think that a species-egalitarian ETI would seek to help whereas an ETI that only intrinsically values itself would seek to harm. However, that is not necessarily true.

A species-egalitarian ETI may not seek to help humanity. It could just leave humanity alone, concluding that helping us would not be worth the effort. This would hold in particular if it does not intrinsically value locations on or around Earth, or if it does not intrinsically value any aspect of humanity. Indeed, an ETI that favors abiotic systems may even seek to destroy humanity and the rest of life on Earth. Alternatively, an ETI that favors ecosystem flourishing could seek to destroy humanity to restore Earth ecosystems or to install other ecosystems on Earth. Finally, an ETI that favors subjective experience or preference satisfaction could seek to destroy humanity because humans are not as capable at having positive subjective experiences or satisfied preferences. Just as a species-egalitarian could value the life of a human more than that of a chicken, so too could it value the life of an ETI more than that of a human.

A species-egalitarian humanity could seek to destroy the ETI for the same sorts of reasons. Alternatively, either civilization could seek to destroy itself upon encounter with the other. For example, suppose humanity learns that the ETI is much more capable of having positive subjective experience than humans can. Humanity could destroy itself in order to free up space in the universe for the ETI to have more positive subjective experience. Now suppose that the ETI intrinsically value ecosystem flourishing, and suppose that it concludes that humanity is better at maintaining flourishing ecosystems. In this case, both civilizations would want to destroy themselves. A race to be the first to self-destruct could ensue (Baum 2010b). This may sound absurd, but it is a logical consequence of certain basic consequentialist frameworks. Furthermore, upon closer inspection it may not seem so absurd, as it is just a particular case of sacrifice for the greater good.

An ETI that only intrinsically values itself may likewise not seek to harm humanity. Again, it could just leave humanity alone. Alternatively, it could find pleasure in humanity's ongoing existence. This is seen for example in the zoo hypothesis scenario, in which the ETI watch humans just as humans take watch nonhuman animals in zoos. A self-valuing ETI is less likely to instrumentally value humans for our capabilities because we are likely to be so much less advanced. We should not count on being able to trade with ETI, just as chickens cannot count on trading with us. Finally, the ETI could keep humans alive to eat us, though distant planets are a long way to go for a new source of food, and our chirality might not be compatible with theirs (Cockell and Lee 2002). Whether this would be a harm to humanity depends on how they would treat us as livestock. While humanity's treatment of livestock is not encouraging, that does not necessarily mean the ETI would behave similarly.

These scenarios are illustrative of the range of ways an ETI encounter could play out. They show how sensitive the outcome is on the ETI's ethics, and potentially also on humanity's. Studies of ETI encounter should have ethics front and center.

8.4.2 Practical Implications

Humanity is not currently in contact with any ETI. However, ETI contact scenarios are nonetheless relevant to at least two types of current human activities: space exploration and treatment of nonhuman animals.

For space exploration, the key message is caution. An ETI encounter could bring major consequences, but those consequences could be either good or bad. That is, they could bring either a large increase or a large decrease in intrinsic value. Humanity should seek the increases and avoid the decreases. Unless we know which is which, we should try to learn more before taking major actions. This point holds across the range of different consequentialist frameworks.

One contemporary decision in which caution is warranted is messaging to extraterrestrials (METI). Sufficiently powerful electromagnetic radiation from Earth can be detected from other points in the galaxy. This includes (1) background leakage from radio and television transmissions intended for Earth audiences, which can unintentionally reach ETI, and (2) messages intentionally sent to ETI. The term METI typically refers to intentional messages.

The consequences of METI depend on signal strength and duration. Stronger and longer signals are more likely to be received by ETI and result in an encounter. Strong signals stand out more from background radiation. Long duration signals, usually done as a repeated signal, are less likely to be interpreted by ETI as an equipment glitch or other false alarm. A sufficiently low-strength, short-duration METI message blends in with background leakage and does not appreciably change the likelihood of ETI encounter (Haqq-Misra et al. 2013). Any ETI that receives these signals presumably already knows about humanity. The signals are thus unlikely to affect the possibility of ETI encounter. Instead, their value is mainly for education on Earth.

In contrast, high-strength/long-duration METI signals are likely to be received by ETI who would not already be aware of humanity. These signals could lead to ETI encounter and are thus of higher consequence. There is an active debate on the question of whether high-power/long-duration METI should be conducted (Shostak, no date). In consequentialist terms, whether METI should proceed depends on whether an ensuing ETI encounter is more likely to be beneficial or harmful, as well as how beneficial or harmful it would be. However, this is very difficult to know because we now have little understanding of what an ETI would be like.

In this situation, the best course of action in consequentialist terms (for the range of different consequentialist frameworks) is to abstain from high-power/long-duration METI in order to evaluate its merits. There is no urgent reason to start high-power METI. It can wait. Once the signals have been transmitted, we cannot take them back. It is thus worth pausing to try to better understand the merits of high-power/long-duration METI. To be sure, gaining a better understanding may be difficult, since it will have to proceed without evidence from any actual ETI encounter. But some progress may be achievable via philosophical inquiry, or from other astrobiological research, such as the ongoing study of exoplanets. Finally, any educational value of high-power/long-duration METI could, more or less, be achieved by a high-visibility study of its merits. Given how high the stakes of ETI encounter would be, this cautious approach is best.

The second practical implication is for the treatment of nonhuman animals on Earth. Humanity often treats nonhuman animals poorly, and justifies it on grounds of human superiority. It is undoubtedly true that humans are superior in certain regards—for example, no other species are exploring outer space. However, humans are not categorically superior, despite some tendencies to place "humans" and "animals" in different categories. (The reader might note my repeated use of the term "nonhuman animals" to emphasize our common lineage.) Humans may be the most intelligent, the most technologically capable, even the most capable of enjoying positive subjective experience or satisfying preferences. But some other species can do these things too.

Whatever humans can do better than nonhuman animals, ETI may be able to do better than humans. Indeed, ETI are likely able to do these things better, because ETI are likely to be much older as a civilization and thus more advanced than humanity.

8.5 Conclusions

A basic conclusion of this paper is that consequentialists should pay attention to outer space. This is because outer space can be the location of immense consequences (via space colonization) and because outer space scenarios can force us to rethink our consequentialist ethics (via ETI encounter).

Attention to outer space prompts us to recognize the big picture. This holds for consequentialist ethics as much as it does for anything else. Only by thinking through the possibilities of outer space can we understand how our lives could matter in the grand scheme of things. And the fact of the matter is that our lives can matter immensely. We can set the pieces in motion for an immense cosmic civilization. We can help prevent civilization-ending global catastrophe so as to enable future space colonization. And we can determine whether or not to try messaging to ETI.

Should we do these things? Answering this all-important question requires ethics. Therefore, just as consequentialists should pay attention to outer space, so too should outer space analysts pay attention to consequentialism, and indeed to ethics in general. Defensible forms of consequentialism will generally conclude that (1) humanity today should focus on avoiding global catastrophe, (2) space colonization should proceed with caution, but ultimately should proceed at immense scale, and (3) high-power/long-duration METI should not be conducted until more effort is put to assessing whether the consequences are likely to be good.

The ethical arguments and empirical analyses in this paper are quite brief and are not the final word on the subject. I have said little in defense of consequentialism and my preferred form of it. The analyses of space colonization and ETI encounter are likewise at best only approximate and leaving much for future work. Some of it is due to space constraints in this paper, but much of it is due to the fact that the research simply has not yet been performed. Outer space consequentialism could make for a fruitful line of inquiry.

The merits of this line of inquiry are diminished by the conclusion to focus on avoiding global catastrophe. Any global catastrophe would preclude the possibility of future research on all topics, including outer space consequentialism. Likewise, any hopes of resolving the ethical dilemmas and empirical uncertainties depend on us surviving long enough to do the research. An argument can thus be made against any work on outer space in favor of work on the global catastrophic risks. My own view is that work on outer space should be pursued mainly to the extent that it is instrumentally valuable towards reducing the global catastrophic risks. To that end it can be quite instrumentally valuable. Outer space can offer great motivation due to its immense opportunities, and it can be deeply inspirational due to its beauty and wonder and the big-picture perspective it offers. While attention to outer space should not distract humanity from the urgent threats that it faces, some attention is very much worthwhile.

Acknowledgments Jacob Haqq-Misra, Ravi Kumar Kopparapu, and Tony Barrett provided helpful suggestions for the development of this paper. The editors provided helpful feedback on an earlier draft. Any errors or shortcomings are the author's alone.

References

Adams, F. C. (2008). Long-term astrophysical processes. In N. Bostrom & M. M. Ćirković (Eds.), *Global catastrophic risks* (pp. 33–47). Oxford: Oxford University Press.
Adler, M. D., & Posner, E. A. (2000). Implementing cost-benefit analysis when preferences are distorted. *Journal of Legal Studies, 29*, 1105–1147.
Asimov, I. (1979). *A choice of catastrophes: The disasters that threaten our world*. New York: Simon & Schuster.
Barry, B. (1989). Utilitarianism and preference change. *Utilitas, 1*, 278–282.
Baum, S. D. (2009). Cost-benefit analysis of space exploration: Some ethical considerations. *Space Policy, 25*, 75–80.
Baum, S. D. (2010a). Is humanity doomed? Insights from astrobiology. *Sustainability, 2*, 591–603.
Baum, S. D. (2010b). Universalist ethics in extraterrestrial encounter. *Acta Astronautica, 66*, 617–623.
Baum, D. (2012). Value typology in cost-benefit analysis. *Environmental Values, 21*, 499–524.
Baum, S. D., Haqq-Misra, J. D., & Domagal-Goldman, S. D. (2011). Would contact with extraterrestrials benefit or harm humanity? A scenario analysis. *Acta Astronautica, 68*, 2114–2129.
Bradley, B. (1998). Extrinsic value. *Philosophical Studies, 91*, 109–126.
Bostrom, N. (2003). Astronomical waste: The opportunity cost of delayed technological development. *Utilitas, 15*, 308–314.
Ćirković, M. M. (2002). Cosmological forecast and its practical significance. *Journal of Evolution and Technology, 12*.
Cockell, C. S. (2005). Planetary protection: A microbial ethics approach. *Space Policy, 21*, 287–292.
Cockell, C. S. (2007). Originism: Ethics and extraterrestrial life. *Journal of the British Interplanetary Society, 60*, 147–153.
Cockell, C. S., & Lee, M. (2002). Interstellar predation. *Journal of the British Interplanetary Society, 55*, 8–20.
Cowen, T., & Parfit, D. (1992). Against the social discount rate. In P. Laslett & J. Fishkin (Eds.), *Justice between age groups and generations* (pp. 144–161). New Haven: Yale University Press.
Diamond, J. (1999). To whom it may concern. *New York Times Magazine, 5*, 68–71.
Drexler, K. E. (2013). *Radical abundance: How a revolution in nanotechnology will change civilization*. New York: PublicAffairs.
Dyson, F. J. (1960). Search for artificial stellar sources of infra-red radiation. *Science, 131*(3414), 1667–1668.
Fogg, M. J. (2000). The ethical dimensions of space settlement. *Space Policy, 16*, 205–211.

Glover, J. (Ed.). (1990). *Utilitarianism and its critics*. New York: Macmillan.
Haqq-Misra, J. (2012). An ecological compass for planetary engineering. *Astrobiology, 12*, 985–997.
Haqq-Misra, J., Busch, M. W., Som, S. M., & Baum, S. D. (2013). The benefits and harm of transmitting into space. *Space Policy, 29*, 40–48.
Holbrook, D. (1997). The consequentialistic side of environmental ethics. *Environmental Values, 6*, 87–96.
Kaku, M. (2005). *Parallel worlds: The science of alternative universes and our future in the cosmos*. London: Penguin.
Michaud, M. A. G. (2007). *Contact with alien civilizations: Our hopes and fears about encountering extraterrestrials*. New York: Copernicus.
Milligan, T. (2015). *Nobody owns the moon: The ethics of space exploitation*. Jefferson, North Carolina: McFarland & Company.
Ng, Y.-K. (1991). Should we be very cautious or extremely cautious on measures that may involve our destruction? *Social Choice and Welfare, 8*, 79–88.
Ng, Y.-K. (2003). From preference to happiness: Towards a more complete welfare economics. *Social Choice and Welfare, 20*, 307–350.
Niemi, G. J., & McDonald, M. E. (2004). Application of ecological indicators. *Annual Review of Ecology Evolution and Systematics, 35*, 89–111.
Nordhaus, W. D. (2007). A review of the stern review on the economics of global warming. *Journal of Economic Literature, 45*, 686–702.
O'Malley-James, J. T., Cockell, C. S., Greaves, J. S., & Raven, J. A. (2014). Swansong biospheres II: The final signs of life on terrestrial planets near the end of their habitable lifetimes. *International Journal of Astrobiology, 13*, 229–243.
Perrings, C., & Hannon, B. (2001). An introduction to spatial discounting. *Journal of Regional Science, 41*, 23–28.
Petigura, E. A., Howard, A. W., & Marcy, G. W. (2013). Prevalence of Earth-size planets orbiting sun-like stars. *Proceedings of the National Academy of Sciences, 110*, 19273–19278.
Portney, P., & Weyant, J. (Eds.). (1999). *Discounting and intergenerational equity*. Washington, DC: Resources for the Future.
Price, C. (1993). *Time, discounting and value*. Oxford: Blackwell.
Rolston, H., III. (1986). The preservation of natural value in the solar system. In E. C. Hargrove (Ed.), *Beyond spaceship earth: Environmental ethics and the solar system* (pp. 140–182). San Francisco: Sierra Club Books.
Rønnow-Rasmussen, T., & Zimmerman, M. J. (Eds.). (2005). *Recent work on intrinsic value*. Dordrecht: Springer.
Sagan, C., & Newman, W. I. (1983). The solipsist approach to extraterrestrial intelligence. *Quarterly Journal of the Royal Astronomical Society, 24*, 113–121.
Scheffler, S. (1982). *The rejection of consequentialism: A philosophical investigation of the considerations underlying rival moral conceptions*. Oxford: Oxford University Press.
Scheffler, S. (Ed.). (1988). *Consequentialism and its critics*. Oxford: Oxford University Press.
Seligman, M. (2010). *Flourish: Positive psychology and positive interventions*. The Tanner Lectures on Human Values, University of Michigan, October 7, 2010.
Shostak, S. (no date). Should we broadcast? And what if we find life in the solar system? SETI Institute, http://www.seti.org/seti-institute/should-we-broadcast-and-what-if-life-in-solar-system
Smith, D. M. (1998). How far should we care? On the spatial scope of beneficence. *Progress in Human Geography, 22*, 15–38.
Stern, N. H. (2006). *Stern review: The economics of climate change*. Cambridge, UK: Cambridge University Press.
Tännsjö, T. (1998). *Hedonistic Utilitarianism*. Edinburgh: Edinburgh University Press.
Tonn, B. E. (1999). Transcending oblivion. *Futures, 31*, 351–359.
Tonn, B. E. (2002). Distant futures and the environment. *Futures, 34*, 117–132.

Chapter 9
Space Ethics Without Foundations

Tony Milligan

9.1 Ethics for Other Places

It seems plausible to say that, wherever we go and whatever we do, respect for persons (and arguably for at least some non-persons) should matter. Irrespective of whether we are here on the Earth or elsewhere, in free space or on another world, there are things that we ought not to do. Murder will be wrong, rape will be wrong and wanton cruelty will be wrong. And this will be true even if we or our descendants are faced with some peculiar fictionally-extreme dilemma which might force our hand and lead us to carry out some dreadful action as the lesser of two evils. Under such circumstances, it would still remain the kind of action about which we ought to feel regret and a sense of guilt, or (in a more classical register) a sense that life has been forever polluted. Perhaps sociopaths and beings utterly unlike ourselves might respond differently, but for people like us these are considerations that are stable over the course of time even though our understanding of what actually constitutes murder, rape and wanton cruelty might change in the future as it has done in the past. No human culture has ever endorsed actions which fall under these descriptions and no stable human culture will ever do so because they are destructive of any sense of community and of the basic levels of trust which needs to exist within a society if it is to survive. To say this is to accept that the cultural relativity of ethics is real but not absolute. Some ethical claims are super-assertable in the sense that they will withstand the challenge of any amount of additional experience, evidence and alteration in our circumstances.

T. Milligan (✉)
Department of Theology and Religious Studies, King's College London,
Room 3.42 Virginia Woolf Building, 22 Kingsway, London WC2B 6LE, UK
e-mail: anthony.milligan@kcl.ac.uk

J.S.J. Schwartz and T. Milligan (eds.), *The Ethics of Space Exploration*,
Space and Society, DOI 10.1007/978-3-319-39827-3_9

Considerations of this sort may tempt us to posit a fundamental level of ethical commitments, a level which is natural or biologically rooted or grounded in social necessities, i.e. baseline claims from which (with the addition of some further data, rules of inference and transitory local norms) a larger body of ethical theory for life in space might be then constructed. And such a move *would* warrant the introduction of a metaphor of ethical foundations. Yet this is a temptation that we have three good (and interrelated) reasons to resist. *Firstly*, the foundations of space ethics would presumably have to be composed of some set of basic (fundamental) principles which would play a major role in ethical deliberation about space and in knowing what to do. The principles in question would have to be both *stable* (such that they remain continuously assertable by the best-placed agents no matter where and no matter when) and also *salient* (such that they are not only true but also relevant to routine ethical practices and decision making). Principles which have only one of these features will be unfit for the task.

A problem that arises from such an approach is that it is by no means obvious that ethical principles are anywhere near as important as this model of ethical knowledge seems to suppose. It is clear that they play an important role in the teaching of ethics; in comparisons of the ethical outlook of one culture with that of another; and as hints and reminders about what might be important across a range of situations. But ethical principles tend to be subject to endless qualifications or else they are so general that they lose the ability to be routinely action-guiding. And this is a comment about the nature of ethics in general rather than anything specifically to do with moving out into space. Foundations for space ethics would require ethical principles to play a role that they don't play in regular terrestrial ethics. (At least, on our best understanding of how ethics works.)

Secondly, there is a more space-focused consideration: based on an understanding of how ethics has developed up to this point (as evidenced by historical shifts in our grasp of exactly what murder, rape and wanton cruelty involve), it is intuitively plausible to say that the sort of ethics that guides people's lives is likely to continue to change in significant ways over the course of time and it is likely to change in significant ways if we ever do establish a society somewhere other than the Earth. This gives us a strong reason to be cautious about setting down any detailed code for a future elsewhere. The difficulty would then seem to be that while certain of the broader contours of ethics are likely to remain stable, a good deal of the detail may well shift. It is not then obvious that the fixed part would be up to the task of serving as foundations. It might be better to think of it as a series of side-constraints which must be met but which do not, on their own, perform much work.

Finally, the comments above address matters of an extremely long range sort: space ethics in the context of a settled life elsewhere rather than space ethics in the more immediate sense of deliberation which tackles the legitimacy of asteroid mining, lunar property rights or whatever astronauts ought to be told by command centers prior to being sent into space. For questions of the latter sort, our existing ethical resources may well be adequate or at least as adequate as they are for terrestrial matters and so for these issues we simply do not need foundations for a special sort of 'space ethic' but rather 'space ethics' thought of as an extension of our regular ethics. Comparisons may be made and dissimilarities drawn between property rights in space and similar rights on the ocean floor; issues within environmental ethics may inform questions about asteroid and lunar mining and about the case for microbial value; and an understanding of risk as a matter of justice because costs and benefits might be shared differently, can be brought into play. These keep us *close to* if not exactly *in* familiar ethical territory.

For such extensions of regular ethics, and even for our initial formulations of an ethics for settlement (i.e. a frontier ethic in the proper sense) I have suggested elsewhere (along with James Schwartz) that some manner of Rawlsian deliberation may be a good place to start (Milligan 2015b; Schwartz 2015). That is to say, human activities in nearby regions of space should be carried out in a just manner and in practice this means that inequalities in the distribution of limited resources such as orbital niches and mining opportunities within the inner solar system, should ordinarily be beneficial to those who are less well-off.[1] Given inequalities of access, and the absence of enough accessible resources to go around, some forms of licensing and compensation systems may therefore be required and standards for sustainable development may need to be put in place to ensure that we do not monopolize resources leaving little for others and for future generations.

We may have to move beyond this in order to construct adequate standards for planetary protection but for many purposes of resource allocation in space, a Rawlsian approach may be adequate if not entirely ideal. It is a framework which I have reservations about in other contexts, but it is at least a framework which is anti-foundationalist. Rather than deduced from foundational principles plus ancillary claims and data, it is reliant instead upon a process of 'reflective equilibrium' in which considered judgements about particular situations and basic principles shape our understanding of which theories are plausible and the latter, in turn, do the same for the principles and considered judgements. Each helps to continuously modify the other, giving us a more adequate picture of what it is like to actually engage in ethical theory and to revise our understanding of the world. While there are some more stable and some less stable assumptions at any given time, all are in principle revisable in a way that foundations would not be.

When it comes to more medium-term issues, such as establishing sizeable and stable human settlements elsewhere in the solar system, our deliberation might also benefit from a familiar kind of Rawlsian thought experiment in which we consider what political and ethical principles we might happen to agree upon if we happened to discuss matters behind a 'veil of ignorance,' i.e. with our rationality and a reasonable self-interest intact but without any knowledge of the ethical or religious traditions to which we happen to adhere, and without knowledge of where we

[1]For the limited nature of the resources which are likely to be within our reach anytime soon, see Ian Crawford (2015) on lunar Helium 3 (^{3}He); Elvis (2014) on the limited number of metallic (m-class) asteroids within our reach; and Schwartz (2015) for the limited available geostationary orbital niches.

personally might be situated in the resulting society (Rawls 1971). If we agree to severe poverty, we might turn out to be among the poor, and so a reasonable self-interest will lead us to avert this danger by maximizing the position of the least well-off.

Yet even this is not exactly the way in which it makes sense to deliberate about most matters of justice on the Earth. With regard to our regular life on Earth, I am not a Rawlsian. I am, rather, a normative pluralist who has been exposed to more than his fair share of virtue-theoretical influences with its emphasis upon the importance of moral vision and practical wisdom rather than principles.[2]

As a result, a Rawlsian framework gives more of a role to principles than is ordinarily the case in everyday life. What I am suggesting, however, is that this is a compromise we can, perhaps, live with, given the non-standard nature of the topics covered by space ethics and, crucially, *our lack of any grounding experience concerning ongoing life off-world*. Put bluntly: we currently have to rely upon the artificiality of principles (and something like a Rawlsian framework) but we may not have to do so once extensive experience of off-world activity has been acquired. (At which point the ladder can be kicked away.)

With regard to regular terrestrial ethics, this concession to Rawlsian deliberation based on the veil of ignorance would be neither required nor helpful. After all, here on Earth we already know where we are actually situated in society. We know whether or not we are wealthy or poor or an adherent to belief system *b* rather than religion *r*. This will always skew our deliberations about what would be agreed to by reasonable but fictional agents deliberating in ignorance of the likely impact of decisions upon their own personal social standing. With regard to settlement elsewhere this is not the case, or at least it is not the case to the same degree. It is, for example, entirely conceivable that a wealthy terrestrial agent might have to surrender their material advantages in order to become part of a settlement program elsewhere. And, if the volunteer lists for implausible settlement plans such as Mars One are anything to go by, there would be no shortage of those who would be willing to make any material sacrifice required in order to secure their passage.[3]

And if we do go Rawlsian about matters a little further afield in addition to appealing to such a framework in relation to the development of a near-Earth space economy, we can make some very plausible claims, e.g. it is extremely unlikely that any reasonable agents deliberating about a planned settlement (as opposed to some incremental laissez faire process) would be keen to transplant all of the inequalities of the Earth into contexts where they could be socially destructive. In the interests of sheer survival, settlements under conditions of extreme vulnerability would

[2]This brings me in line with some aspects of the thinking of Sparrow (1999, 2015) although I tend to be less agent-based and less skeptical about the moral defensibility of human activity in space. For a critique of this virtue theoretical approach, see Jim Schwartz' contribution to this volume.

[3]For the impracticality of the Mars One plan, even given some fairly generous assumptions, see the MIT feasibility study (Do et al. 2014).

require a strong sense of social solidarity that radical inequalities might undermine. If they could be avoided, without all of the dangers of unraveling inequality on Earth, it seems plausible to say that behind a veil of ignorance, without a baseline reassurance that we would remain personally wealthy, many well-informed agents would not take the risk. For better or worse, the kind of political and ethical arrangements which can secure social stability, solidarity and the conditions for a good life *anywhere*, will often need to be shaped by more than the bare information than Rawls allows us to take behind the veil (Milligan 2015b). But for certain matters of near term justice in space, and for medium term space settlement, the non-foundational Rawlsian approach (and the associated 'veil of ignorance' thought experiment) work well.

9.2 Ethical Change and Truth

Understood thus, the Rawlsian approach occupies a narrow band-width of relevance. Beyond these limits, it is tempting to revert to some form of naturalism in order to delineate the bounds of what might be ethically viable and, up to a point, this may make sense. However, it will do so only up to a point. Our successors elsewhere, even within the solar system, are unlikely to remain biologically indistinguishable from ourselves for an indefinite period of time. While they may remain members of our 'human' community in other and non-biological senses, what would become 'natural' in an extremely unfamiliar environment is not at all obvious. Nor do we really know how quickly the process of differentiation would occur. (Rapidly enough, perhaps, to make good lives a really possibility in otherwise unforgiving environments.)

Yet, even within its narrow range of applicability, I have suggested elsewhere that the Rawlsian device of the veil of ignorance might yield some surprising results, and results which may also reinforce our sense of the relativity of the ethical. On the surprising, and surely welcome, side it might yield a defense of various kinds of constrained dissent as a safeguard against the emergence of authoritarian systems towards which the vulnerabilities of space would create considerable pressures. (A point made forcefully in Cockell 2013, 2015.) An absence of a local culture of constrained dissent would risk generating far more dangerous, if more intermittent, forms of dissent foreshadowed in science fiction, and thereby system collapse. Up to a point, a liberal ethic much like our own may turn out to be viable under conditions of settlement, but not necessitated by the latter, and (again *up to a point*) some form of democracy may actually be the most stable long term political option in the kinds of 'human-friendly' places to which we might initially go. Although, as an important qualification, it is unlikely to look like the kind of democracy that we are familiar with. These are welcome thoughts for many reasons, not the least of which is that they line up well with respect for persons as a basic side-constraint. Without this, settlement of any sort would be extremely difficult to justify.

What is less welcome is that some familiar and important terrestrial freedoms might be much harder to transfer. It is not obvious that settlements of several hundred or several thousand humans could tolerate agents simply opting out of the labor process, whether through wealth (thereby becoming free-loaders) or through the pursuit of some alternative lifestyle which might not constitute a steady drain upon resources but would nonetheless involve an initial loss and ongoing missed opportunities. Our conception of what constitutes freedom with regard to work might have to be situation specific and the penalties for persistent non-cooperation with fellow citizens, even if they did not exactly amount to pushing someone into an airlock, could strike us as draconian if applied here on Earth. Similarly, and from a broadly liberal standpoint rather more alarmingly, abortion rights might well be subject to additional constraints. Agents deliberating behind a veil of ignorance about their own likely position within society, but with basic knowledge of where a settlement was to be located, its impact upon human reproduction and a concern to order a sustainable society, might have to accept that abortion (except for certain kinds of medical reasons) should be restricted, penalized or even banned. Given the extent to which altered gravity and microgravity can affect the capacity to conceive and carry to term, it is not clear that any sustainable society could afford to regard reproduction as strictly a matter of individual choice rather than a social priority.

This should not be read as, in any way, a case of ‘smuggling,’ the convenient encoding of a pre-determined terrestrial pro-life/anti-abortion position within space ethics, given that (i) the same considerations of reproductive difficulty manifestly do not apply here on Earth; and (ii) parallel consideration might necessitate the *promotion* of abortion under conditions where reproductive problems were limited but where population levels threaten to outstrip the carrying capacity of the available infrastructure. It is extremely unlikely that population matters could be approached in the laissez faire manner that underpins a good deal of opposition to abortion rights. Population might have to be managed, at least until mechanisms for equilibrium were in place. Nonetheless, the considerations in play here outstrip the bounds of a broadly liberal ethic and conflict with familiar intuitions about bodily integrity and autonomy (which reach deeply into our understanding of issues in bioethics).

Elsewhere, I have suggested that considerations of this sort might reasonably impact upon any future decisions about where we attempt to settle (Milligan 2015b, 2016). If local conditions would place too many pressures upon our familiar rights in order to establish a stable settlement then perhaps we should not go. There is, after all, little to be gained from the ‘Iron Skies’ option of helping to establish an illiberal or even totalitarian system simply in order to maximize the number of settlement locations. However, imperfect knowledge and the extent to which guess-work will be involved in any such deliberations could mean that somewhere, at some point in time, tough decisions might have to be made and familiar liberties or ethical commitments might have to be limited, suspended, modified or jettisoned. And here we begin to see how the particularity of circumstance starts to erode any sense that we could always be guaranteed to live elsewhere as we do now, by appeal to any reassuring set of foundational considerations.

Admittedly, thoughts of this sort raise several causes for concern, not the least of which is that ethics may begin to look so flexible that any notion of it being a matter of truth rather than expediency is in danger of being lost. Yet this does not do justice to either the particularity or situation-relative component of ethics. There is a significant distinction between speaking about ethics as relative and claiming that it is merely subjective. To accept that it is right to do x in situation 1 but wrong to do x in situation 2 is not at all to deny the possibility of ethical truth. It is consistent with claiming that it *really would be right* to do x in situation 1 but *really would be wrong be wrong* to do it in situation 2. Just as it really is (sometimes) ethically permissible to sing raucous songs at sports events but not permissible to do so at funerals or the middle of an academic examination (all other things being equal).

Similarly, as a general principle, we ought to tell the truth, but we should not do so to a well-armed madman who asks directions in pursuit of his would-be victims; nor should we tell the truth in ways and at points in time which are likely to lead to various sorts of easily avoidable harm. If, out of sheer malice, I buy the right to inform someone that their partner is guilty of infidelity I can hardly claim to be acting out of a virtue of truthfulness (or virtue of *any* sort). The point here is that the complex requirements of being truthful go beyond the limited guidance which is offered by principles or rules, although the latter may serve as guidelines of a sort. This complexity of ethics requires a situational sensitivity that is constrained but is rather different from mere expediency. What I am arguing for here is situational sensitivity of a more stable sort. More stable because it concerns circumstances which are likely to prevail for extended periods of time but will differ from place to place rather than merely occasion to occasion.

What is nonetheless presupposed here is the idea that while the right thing to do might be different in a settlement and on Earth, there genuinely are disciplined standards for ethical discourse, and more particularly that claims in ethics have genuine truthmakers, i.e. there are non-arbitrary considerations that make the claims in question true even if their truth is indexed to particular contexts. However, they are true in ways which resemble the ways that various claims about art or aesthetic matters or comedy are also true, i.e. in ways which depend upon how certain kinds of authoritative agents would respond to the matters in question and why they would do so. Truth in these domains is, accordingly, very different from truth in mathematics where characteristic patterns of agent responsiveness is not relevant to fixing what is true and what is false. There are also respects in which moral truths differ from those of comedy and art, their motivational force being one obvious example.

Note also that this way of cashing out matters, in terms of what is familiarly called a 'response-dependence' approach, avoids some of the oddities and arbitrariness of familiar attempts to base ethics solely upon judgements of what has, and what does not have, 'inherent value.' Discussions of the latter sort tend to evoke two sorts of reactions from those who work in the sciences: (a) dismissal as a fiction because such value is not the kind of thing which would appear in a complete inventory of 'what there is' compiled by the best conceivable science; (b) a 'pick a property…any property' approach which fasts upon some particular property or properties (rationality and sentience are favorites) which are taken to be the key to value and which just so happen to line up with current practices of valuing and/or offer conveniently few barriers to the conduct of science.

I am not, when I say this, denying the importance of claims about inherent value. Indeed I am fairly robust in my understanding of value bearers. What has made life difficult in this context is, rather, that we have become tangled up in this otherwise useful part of our conceptual equipment. To disentangle matters, such talk is often a kind of shorthand for a story about the reasons that we or other agents have for acting and responding. In a large class of cases such talk is translatable. So, to say that *O* is inherently valuable is (in a large class of cases) much the same as saying that we have *reasons for acting* which do not concern anything other than the flourishing of *O* or the avoiding of harm to *O*. A story is then due about what reasons for action might be, and different candidate stories will no doubt be offered, but whatever we say about such reasons, it may make sense to say that they can be situation sensitive. That is to say, we may have reasons for doing one thing in one situation and something else in another situation. Or in one situation our reasons may be stronger than in another. This is the way in which we ordinarily understand what it is to have a reason, even when we are not engaging in ethics. And a familiar example when we are engaged in the latter is the issue of microbial value: how we might have reasons for microbial protection on other worlds which we do not seem to have on this one. And this is *not* the sort of question which we can answer by comparing structure and finding that extraterrestrial microbes have a magical property underpinning special value which their poorer terrestrial counterparts lack. Rather, our reasons for treating them differently will depend upon the situation and context in which they exist. (Another way of putting the point would be to say that they have *relational* properties of an important sort.)

9.3 The Scope of Space Ethics

If the truth of claims about ethics (including space ethics) is in some way bound up with how authoritative agents might respond, what follows? Well, at least a certain difficulty. Whereas we ordinarily have a good grasp of who might be an authoritative agent with regard to terrestrial matters, and how such agents might see, and respond to, particular actions and events, we have very little grasp of who might count as an authoritative agent in the context of distant space settlements with a range of vulnerabilities and psychological pressures that we simply cannot appreciate. Understanding what it is to be such an agent is very different from understanding a science-fiction film. Beyond accepting various important platitudes about murder, rape and cruelty (a good number of which any livable ethic would have to satisfy) we simply have very little idea of how they would see their worlds and respond to them, very little sense of what the best sort of response would involve.

What this means is that the content of any account of the foundations of ethics which we could actually specify, and which might be shared between ourselves and such future agents, would have to be exceptionally thin. It would have to be a list of platitudes in the strict sense, i.e. claims of such an extremely general sort that we are all likely to affirm (and which might establish minimal adequacy conditions for a plausible ethic) but from which very little can actually be deduced. Platitudes of this sort can be very useful. We probably cannot do without them. They may certainly help us to tell if we are on roughly the right track, but they may not do much more than this sort of odd-job. As foundations in the initially specified sense of being both *stable* (in terms of truth or even assertability by authoritative agents at all times and in all places) and also *salient* to the deduction of detailed ethical judgements when considered in conjunction with various more local items of knowledge, they will simply be unfit for the task. Stability they might achieve but not stability *and* deductive salience. Indeed, the former will arguably be secured only at the expense of the latter. The more general they are the more stable they will be, but also the less informative.

If this is right then something noteworthy follows about the proper scope of space ethics. Arguably, it has an important role in shaping our deliberations in near and medium-term contexts, the requirements of justice within the latter, what would constitute a sustainable program of activities in space and the rudimentary shape of an appropriate contemporary attitude towards space as the next frontier. This will probably cover us adequately for discussions about the ethics of early settlement but not far beyond it. At some point the frontier simply turns into more of an event horizon. Or, at least, it is not obvious that space ethics is going to be the most illuminating sort of discourse that we can currently bring to bear upon matters in the more distant future. When we attempt to stretch the discourse beyond such bounds it may remain edifying but it will begin to resemble a form of science fiction or perhaps even to constitute a form of the latter. We might also expect it to win fewer awards. (Kim Stanley Robinson and Stephen Baxter need not feel threatened.) In any case, the line between the two will become blurred and it will do so for a good reason.

This is not necessarily a bad thing, but it does mean that the reliability of familiar sorts of ethical deliberation in such contexts will either be compromised or else they will function as a coded way of commenting upon the present or at least upon more proximate matters. (About which something more reliable *can* be said.) Here, I am drawn to think of my own faltering attempts to make sense of the ethics of life on a multigenerational ship, *en route* to some other star system, and how indispensable it was to couch the discussion in terms of various classic science fiction treatments of the scenario rather than to build it out of fundamental principles (Milligan 2015a, pp. 134–51). To be sure, something useful can be said in such discussions, but anything deep that is said may turn out to concern our current predicament or what it is to be human rather than space exploration as such. And this is slightly paradoxical because it means that the further we try to reach into a human future in space and understand the ethics of such a future, the more we are thrown back upon what is familiar, proximate and deep. The danger then is one of imagining that we can specify, by appeal to known and homely considerations, more than the very broadest and most general ethical features of how this more distant and troubling future might be lived.

References

Cockell, C. (2013). *Extraterrestrial liberty: An enquiry into the nature and causes of tyrranical government beyond the earth*. Edinburgh: Shoving Leopard.

Cockell, C. (2015). Freedom in a box: Paradoxes in the structure of extraterrestrial liberty. In C. Cockell (Ed.), *The meaning of liberty beyond earth* (pp. 47–68). Heidelberg, London and New York: Springer.

Crawford, I. (2015). Lunar resources: A review. *Progress in Physical Geography, 39*, 137–167.

Do, S., Ho, K., Shreiner, S., Owens, A., & de Weck, O. (2014). An independent assessment of the technical feasibility of the mars one mission plan. In *65th International Astronautical Congress, Toronto, Canada*. IAC-14-A5.2.7.

Elvis, M. (2014). How many ore-bearing asteroids? *Planetary and Space Science, 91*, 20–26.

Impey, C., Spitz, A. H., & Stoeger, W. (Eds.). (2013). *Encountering life in the universe: Ethical foundations and social implications of astrobiology*. Tucson: University of Arizona Press.

Milligan, T. (2015a). *Nobody owns the moon: The ethics of space exploitation*. North Carolina: McFarland.

Milligan, T. (2015b). Rawlsian deliberation about space settlement. In C. Cockell (Ed.), *The meaning of liberty beyond earth* (pp. 9–22). Heidelberg, London and New York: Springer.

Milligan, T. (2016). Constrained dissent and the rights of future generations. In C. Cockell (Ed.), *Dissent, revolution and liberty beyond earth*. Heidelberg, New York and London: Springer.

Rawls, J. (1971). *A theory of justice*. Cambridge: Harvard University Press.

Schwartz, J. S. J. (2015). Fairness as a moral grounding for space policy. In C. Cockell (Ed.), *The meaning of liberty beyond earth* (pp. 69–89). Heidelberg, London and New York: Springer.

Smith, K. C. (2014). Manifest complexity: A foundational ethic for astrobiology? *Space Policy, 30*, 209–214.

Sparrow, R. (1999). The ethics of terraforming. *Environmental Ethics, 21*, 227–245.

Sparrow, R. (2015). Terraforming, vandalism, and virtue ethics. In J. Galliott (Ed.), *Commercial space exploration: Ethics, policy and governance* (pp. 161–178).

Part III
Humanism and Posthumanism

Chapter 10
Why Space Migration Must Be Posthuman

Francesca Ferrando

10.1 Introduction

What would you do if you won a free ticket to Mars? And what if, after the initial excitement, you realized that it is a one-way ticket? Soviet cosmonaut Valentina Tereshkova, who in 1963 became the first woman to go to space, would most likely be thrilled. In 2013, she offered to volunteer for a one-way trip to Mars, declaring, at age 76: "We know the human limits. And for us this remains a dream. Most likely the first flight will be one way. But I am ready" (Smyth 2013). This chapter will reflect on the "human limits" mentioned by Tereshkova in the context of space migration, addressing them not only from a technological and biological perspective, but also from an ethical, socio-cultural and onto-epistemological one. Such an analysis will lead to emphasize the urgency of adopting a posthuman approach in the process of space-migration, which will be engaged upon not just as a speculative hypothesis, but as an actual possibility. Following the current activities of space exploration, space commercialization and space tourism, NASA stated that the capability to send humans to an asteroid will be developed by 2025, and to Mars in the 2030s (NASA's journey to Mars, n.year); private companies such as SpaceX are setting even closer deadlines, such as 2026 (Hooton 2014), while the non-profit foundation Mars One is planning to establish the first human settlement on Mars live on a TV show.[1] Even though the technical feasibility of the project is considered scientifically unsustainable by many (Do et al. 2014), this chapter will take

[1]This is the plan, as stated in the Mars One website: "Crews will depart for their one-way journey to Mars starting in 2026; subsequent crews will depart every 26 months after the initial crew has left for Mars" (Human Settlement on Mars, n. year).

F. Ferrando (✉)
Liberal Studies Program, New York University, New York, USA
e-mail: ff32@nyu.edu

J.S.J. Schwartz and T. Milligan (eds.), *The Ethics of Space Exploration*,
Space and Society, DOI 10.1007/978-3-319-39827-3_10

into account the enthusiasm that Mars One's vision was met by: thousands[2] of volunteers from over 140 countries (Over 200,000 apply, n. year) applied for a one-way ticket to Mars. They are ready. Tereshkova is ready. What about humans, as a species... are we ready?

Etymologically, the term "human" comes from the Latin term "humus"[3] meaning "soil", which, in our solar system, is only present on Earth. We can thus see migrating to space as the linguistic and semiotic step towards the literal creation of post-humans, that is, beings "post" (Latin for "behind" and "after") their earthly provenance. Furthermore, as we will see in the course of this chapter, space migration will expand the notion of the human, aligning it with a posthumanist sensitivity. In the history of planet Earth, most human societies have developed around dualistic ways of thinking, based on symbolic binaries such as: human/robot, human animals/non-human animals, female/male, black/white, good/evil, nature/culture, self/other. Such a dualistic mindset brought along bio-centric, human-centric, sexist, racist, ethnocentric practices and homophobia, along with eco-disasters and war. If humans migrate to space with a dualistic mindset, and if history is any indication, "space colonization" is then likely to precipitate species discrimination and planetary wars.

Language is not innocent: in order to set a post-dualistic approach to our futures, we should start with a critical analysis of our own terminology. The postmodern post-colonial legacy of the posthuman does not support the use of the term "space colonization", since the notion of "colonialism" is embedded in historical contexts and discriminatory policies which have been rigorously analyzed and criticized within the field of Post-Colonial Studies (cf. Said 1978; Spivak 1987). This chapter will adopt, instead, the term "space migration" , offering a revisitation of humanistic, anthropocentric and Earth-centric practices. And still, space cannot be analyzed in separation from Earth: these realms are inextricably related and shall be investigated in conjunction. In order to demonstrate this important point, we will reflect upon the relevance of the study of celestial bodies in the formation of human civilizations; then, we will highlight the impact of current space technology on planet Earth; thirdly, we will delve into the relevance of space migration to a revision of the notion of the human itself. Posthumanism, as a post-humanism (in the sense of the humanistic tradition), a post-anthropocentrism (Braidotti 2013) and, more in general, a post-dualism, represents a well suited philosophy to pursue this onto-epistemological shift. The dynamics of space migration will thus be inquired

[2]The precise number of applications is being disputed, varying from more than 2000 (Keep 2015) to more than 200,000, according to the Mars One website.

[3]Is important to note that this etymology is not unanimously accepted and that it has been contested for different reasons. From a linguistic perspective, it has been noted the change of the vowel "u", which in "humus" is long (ū), while in "humanus" becomes short (ŭ) (cf. Romaniello 2004, 188–190). This interpretation has also been challenged for its semantics. As early as the 1st Century AC, Marcus Fabius Quintilianus (c. 35–c. 100) stated (Institutio Oratoria I, 6, 34): "Are we to assent to the view that *homo* is derived from *humus*, because man sprang from the earth, as though all other living things had not the same origin...?" (Quintilian and Butler 1920, 127).

by reconciling the varied philosophical landscape of the posthuman, bridging different schools of thought such as: Philosophical Posthumanism, Transhumanism, New Materialism and Object-Oriented Ontology.

10.2 Children of the Stars: *Being in Space*

Before accessing the topic of space migration, we shall start our reflection by asking the question: "should we go to space?". A common answer given to this question is that humans should first take care of the problems characterizing planet Earth, before going somewhere else.[4] The counter-argument to this point has been mostly utilitarian, that is: we will never go to space, if we wait until we resolve all the problems on Earth. Or, alternatively: we will resolve problems on Earth by going to space and finding new resources. Instead of an utilitarian view based on a linear progression of time, which has been criticized both from a scientific (Gould 1996) as well as from a philosophical perspective (especially, within the post-modern frame), we will approach this subject from an onto-epistemological standpoint inspired by Heidegger's definition of technology (1953; Engl. Trad. 1977); space will be thus accessed as "a way of revealing" (ivi, 12), allowing for an original understanding of the notion of space. Let's start with the problematics related to our initial question by asserting a basic and still informative point: we have always been in space. Humans have originated on planet Earth, which is part of a solar system located, among many others, in the Milky Way galaxy. Asking if we should go to space, more than looking for a real answer, reflects the human-centric dualistic attitude of presenting the unknown as a form of separated otherness. A more precise question could then be: "should humans migrate to outer space?". The term "outer space" is defined by the Oxford Dictionary as "the region of space beyond the earth's atmosphere or beyond the solar system. In extended use: a place or region beyond the usual limits of awareness or accessibility" (2015). It can be seen as an extended and somehow confused framework, comprehensive of a wide variety of spaces, such as the solar system as well as the regions beyond the solar system. The notion of "outer space" implies the one of "inner space", which would be the Earth and the earth's atmosphere: these dualistic premises become disputable when realizing that outer space is crucial to the understanding of life on Earth and to the development of the notion of the human itself.

From the origin of human civilizations to the contemporary world, the skies have played a crucial role in the formation of human identity. Astronomy[5] is considered the oldest of natural sciences, dating back to the Upper Paleolithic (cf. Brady 2013); its relevance to many ancient cultures went so far as becoming focal part of their

[4]Saara Reiman calls this approach "Earth first" (2009, 83).

[5]Let's note that the notion of astronomy in the ancient world cannot be fully assimilated to the contemporary scientific field.

social identity and architectonical outfits.[6] The large majorities of the temples across world civilizations were built in accordance with astronomical alignments: think, for instance, of the solstitial orientation of the sarsen monument at Stonehenge (Ruggles 2006) or the early Maya urban planning architectural complexes (Šprajc et al. 2009). The connection between knowledge of the skies and the sacred were deeply intertwined, and it is broadly believed that the astronomers of the ancient times were also the spiritual guides or priests of their communities (cf. Hayden and Villeneuve 2011). In Ancient Egypt, astronomical imagery was an essential part of their religious pantheon, with Ra, the Sun God and the raising sun as symbols of creation; his daughter, Goddess Maat, represented the Order of the Universe and was the wife of the Moon god Thoth. In the words of archaeoastronomer Giulio Magli: "…celestial matters were, for the ancient Egyptians, deeply and intimately connected with the most important things of all: preserving *Maat*, the cosmic order, on Earth. Such an order was *anchored* to the celestial cycles: the cycle of the sun, the calendar, the succession of the hours of the night, the reappearance of Sirius and of the other stars" (2013, 2). The star Sirius, sometimes called "the Dog star" due to its location in the constellation of Canis Major (Latin for the "Greater Dog"), still has a special meaning in the creational myth of the Dogon, an indigenous tribe of southeastern Mali in West Africa (Griaule 1965).

The prehistory and early history of humanity do not treat space as otherness, but as an integral part of the human genealogy, providing knowledge of great relevance for social functioning and daily survival. Similarly, on a scientific level, a strict division between the Earth and the heavens cannot be placed either. Scientific hypotheses such as exogenesis and panspermia[7] postulate that life did not originate on Earth, but formed somewhere else extraterrestrially.[8] A consideration of Earth ejecta and the hypothesis of lithopanspermia, that is, the potential for organic molecules preserved in space as a consequence of "the collision of kilometre-scale bodies with Earth, comets or asteroids" (Reyes-Ruiz et al. 2012, 777), also plays an important role in the disruption of a strict division between Earth and outer space. With these premises, we can now go back to our question: "should humans go to outer space?". Here we shall refresh another simple, yet important and multilayered piece of information: humans have already been in outer space in a series of successful launches achieved as the result of the space race between the Soviet

[6]A specific area of studies called "archaeoastronomy" has recently developed around this topic (cf. Magli 2009).

[7]Such a gender-unneutral nomenclature is first to be found in the writings of Anaxagoras (c. 510–428 BC). Its Greek etymology of pan ("all") and sperma, which in ancient Greek referred to both "origin, source" and to "(human) seed" (translation by Slater 1969), reflects the sexist, and scientifically inaccurate, vision which identified the active principle of life in the male reproductive fluid, while the female was considered to contribute passive matter. This view influenced the ways in which models of conception were described in standard Western science, until as late as the 1980s (Cordrick 2008, 69–70).

[8]It is important to note that the question of where life began does not answer the question of how life originated. For further investigation on the origins of life, see among others, Hazen (2005).

Union and the US during the Fifties and Sixties of the twentieth century (cf. Cadbury 2005). The modalities of the space race will be analyzed in order to show how humans have lost in it their ontological primacy.

10.3 Dog Superstars: *Going to Space*

The first animal to orbit the Earth was the Soviet dog Laika in 1957. Her touching life story opens many questions regarding animal rights, space ethics and human entitlement. Laika was a stray dog, which was selected, due to her physical characteristics and her calm personality, to be the occupant of Sputnik II. At that time, the technology to retrieve the spacecraft had not been developed yet, so her death was certain: Laika died within a few hours after launching; the cause and time of her death were made public only in 2002 (cf. Turkina 2014). This mission was followed by a number of other animals flown into space, including monkeys, rabbits, frogs and mice. In 1960 Belka and Strelka, two other stray dogs, were the first Earth beings to make it back from space on the Soviet Sputnik 5.[9] The relevance of non-human others in outer space has been crucial in the history of space exploration: "[a]nimals would prove to be our bridge to the future. They gave substance to the vision. If humans could not yet travel in space, animals could" (Burgess and Dubbs 2007, XLVIII). Human biology has not evolved to survive in outer space, but to adapt to the great variety of conditions found on planet Earth. With few exceptions,[10] current physiological outfits of human and non-human animals are designed to live in gravity: the bones, muscles, fluid shift, cardiovascular system, among other bodily parts and performances, are effected by microgravity (cf. Russomano et al. 2008). The overall consequences of space travel on the human body are numerous and not yet fully predictable. For these reasons, some scientists do not support human migration to space, finding instead an ideal candidate in robots.

Given the wide variety of possibilities when designing robots and their independence from biological needs, machines have proven to be well suited to outer space conditions. As space scientist James Van Allen highlights: "Almost all of the space program's important advances in scientific knowledge have been accomplished by hundreds of robotic spacecraft" (2004). This robotic advantage[11] has brought some scientists to claim that only intelligent machines should be sent to

---

[9]As the head of the recovery team, Arvid Pallo recalls the joyful reaction of the two dogs: "they were visibly pleased to be back on Earth" (Burgess and Dubbs 2007, 206).

[10]The tardigrades, for instance, are tiny invertebrates known to be the first organisms to be found to survive the vacuum of space (cf. Guidetti et al. 2012, 99).

[11]One problem with robots has been their incapacity to adapt in case of injury, although a recent study has successfully employed specific algorithms which allowed robots to find ways to compensate for the damage (Cully et al. 2015).

outer space, instead of humans, but as Launius and McCurdy make clear, this is a "false dichotomy" (2008, IX). When thinking of space, the traditional division human/robot does not apply any more. Going to space can be seen as an actualization of the notion of the assemblage, perceived by Deleuze and Guattari as a "fuzzy aggregate" (1980; Engl. Transl. 1987, 344) of which "[t]he material must be sufficiently deterritorialized to be molecularized and open onto something cosmic" (ibid.). Such an extended identity can be epitomized by the case of the Mars rovers. Out of a number of successful launches, currently two rovers named "Curiosity" and "Opportunity" are conducting searches and studies on Mars. Interestingly, most scientists working on these missions use the first person perspective to describe the work of the robots (Clancey 2012). Specifically, when the scientist uses the first singular person, s/he is referring to the self in assemblage with the rover. For instance, one scientist stated: "Meridiani [the landing site for "Opportunity"] was such a different landing site because *everything was so laid out in front of me*" (ibid., 100, emphasis in original). Alternatively, when the first plural is used, the scientist is referring to the assemblage of the team of experts working from Earth, and the rover on Mars. For instance, another scientist affirmed: "I remember *we* landed at the Gusev site and *we* looked at the Columbia Hills and *we* joked about it, *we* laughed" (ibid., 101, emphasis added). In this post-dualistic frame of bio-machinic assemblages, it is important to underline the fact that not every human has been granted the same opportunity: depending on their race, gender, sexual orientation and so on, some human beings have been discriminated against in the historical practices and narratives of space exploration.

The first man and woman to make it to outer space were both Soviet cosmonauts[12]: Yuri Gagarin (1934–1968) in 1961 and, Valentina Tereshkova (1937–) in 1963. Despite the success of her operation,[13] Tereshkova woul7d remain the first and last woman in space for almost two decades, until 1982, when Svetlana Savitskaya (1948–), another Soviet cosmonaut, became the second woman in space. The third one was Sally Ride (1951–2012), who in 1983 became the first American woman and the youngest American astronaut to travel in space. After her death, Ride was also celebrated as the first known LGBT astronaut.[14] This significant American delay in promoting female astronauts reflects the inner sexism of NASA

[12]Here, I would like to briefly mention that the terms "cosmonaut" and "astronaut" are mostly synonyms: the term "cosmonaut" is used by the Russian Space Agency, while "astronaut" is used by other agencies such as NASA, ESA and JAXA. For a reflection on cultural-specific values of such terms, see Langston and Pell (2015).

[13]During the first day of her solo flight, Tereshkova detected an error in the automatic navigation software: the spacecraft was moving away from planet Earth, instead of going back towards it. Her mission could have turned into a tragedy, but Tereshkova was able to manage the situation. She reported the error and was thus able to save her life and complete the mission (cf. Evans 2009). Note that the error in the automatic system was made public only in 2004 (ibid., 58).

[14]As Garofali reports, "Sally Ride never 'hid'" (2012), she was just very private about her personal life.

at the time.[15] Gender discrimination was paired with race discrimination, as Kim McQuaid reminds: "the U.S. astronaut corps stayed closed to women in the late 1950s and early 1960s…it opened to women of all races and minority men after 1978" (2007, 421). All of the twelve astronauts of the Apollo Program who walked on the Moon were of male gender and of white ethnicity. In 1980, the Cuban cosmonaut Arnaldo Tamayo Méndez (1942–), selected as part of the Soviet Union Interkosmos program, became the first astronaut of African descendent; in 1983, Guion Stewart Bluford (1942–) was the first American of African descent in space. If gender and race are significant categories of reflection in a critical analysis of the history of humans in space, nationality is another important factor to consider. Currently, only a limited amount of countries have space programs; within these countries, some have gained an historical primacy. Russia, for instance, has a long-lasting legacy in space travel, which is still very actual: since 2011, NASA has relied on Roscosmos, the Russian Federal Space Agency, to fly astronauts to space stations. The centrality of Russia in space matters was historically and ideologically supported by the birth of the cosmist movement,[16, 17] in the twentieth century, which would promote, among other themes, "the active human role in human and cosmic evolution; the creation of new life forms, including a new level of humanity" (Young 2012, 4).

Space migration will open not only new markets, new economies and new resources, but it will also open the biological outfit of the human to new evolutionary exposures and adaptations to alien geographies, breeding into the conversation the transhumanist ethical discussion on human enhancement (cf. Bostrom and Roache 2008). This topic unfolds a series of challenging opportunities. For instance, let's consider this hypothetical scenario: if it was scientifically demonstrated that beings created by merging human and non-human DNA were better suited to survive outer space conditions, would the development and creation of bio-technological hybrids and chimeras[18] be ethically acceptable in the social, legal and biopolitical frame of space migration? On one side, the deconstruction of an essentialist notion of the

---

[15]An example of gender discrimination is the failure of the privately funded program later defined as Mercury 13 (1960), in which thirteen women successfully underwent some of the same screening tests used by NASA to select their male astronauts for project Mercury (cf. Ackmann 2003).

[16]Nikolai F. Fedorov (1829–1803), one of the theorists of Russian Cosmism, had a deep influence on Konstantin Tsiolkovsky (1857–1935), whose scientific work largely contributed to the success of the Soviet space program.

[17]It is worth noticing that in Russian the term "cosmos" simply means "space"; following, the terms "cosmonaut" and "cosmism" are related to this notion.

[18]Rebecca Ballard clarifies the difference between hybrids and chimeras: "In a hybrid, each species combined contributes to half of the DNA contained within a single cell resulting in a blending of the two species characteristics and every cell in the body has that same genome. In contrast, a chimera is a product of grafting cells from one entity to another, rather than the blending one genome with another, which creates a mosaic of mis-matched parts because each population of cells retains its own distinct characteristics" (2008, 302). Ballard's article also offers a medical and legal overview on the topic of animal/human hybrids and chimeras.

human may support the ontological acceptance of such beings. On the other side, the adaptation of human biology to outer space may eventually lead towards the evolution of hypothetical posthumanities, that is, future generations of beings evolutionarily related to the human species but no longer definable as such (Roden 2015). The ethical challenges of natural-cultural evolution find an interesting arena in the varied landscape of the posthuman theoretical scenario. According to New Materialism, there is no strict division between nature and culture, as Karen Barad emphasizes: "[t]he relationship between the material and the discursive is one of mutual entailment…Neither discursive practices nor material phenomena are ontologically or epistemologically prior" (2003, 822). By introducing human enhancement into the debate within the frame of adaptability, we shall also address the crucial subject of genetic preservation, since the risk of erasing biological diversity on Earth may result as a long-term consequence of widespread use of genetically modified organisms and biotechnological manipulation (cf. Shiva 1993). In the post-dualistic frame of Philosophical Posthumanism, the past is not separated from the future; planet Earth is related to outer space; humans are an integral part of their ecosystem. Earth heritage is relevant not only to future generations of humans, but also to hypothetical posthumanities, as well as to current and future robots and intelligent machines. In his classic study on Artificial Intelligence (1988), Hans Moravec sharply stated: "it will be in our artificial offspring's power, and to their benefit, to remember almost everything about us" (1). To support the future we have to respect the past. Destroying Earth's heritage will deprive our future generations of their roots, preventing them from looking further into the heavens; instead, they will have to retrieve crucial knowledge lost under the dust.

10.4 Under the Stardust: *Space Technology*

In order to comply with a comprehensive analysis of space migration, after asking "who is going to space?", we shall ask: "How are we going to space?". In this section we will highlight the fact that, although the ideology according to which space has entered the international debate shares a posthumanist sensitivity, its pragmatics are based on anthropocentric and Earth-centric practices which shall be reflected upon and disputed. Let's start by focussing on space technology. Space rockets are products of the historical frame of War World II. In 1944 the V-2 rocket was the first human-made object to cross the Kármán Line[19]; it was developed as a "vengeance weapon" by Nazi Germany[20] against the allies and it was built by

[19]The Kármán Line is considered the boundary between Earth's atmosphere and outer space (cf. Darrin and O'Leary 2009).

[20]It is worth noticing that, on a psychoanalytical level, the phallic iconography of spacecraft (Rabkin 2002) appears symbolically in tune with the "Fatherland" the Nazis were fighting for (cf. Petersen 2009).

prisoners of concentration camps. One of the key figures of the technology behind the V-rockets was Wernher von Braun (1912–1977), an aerospace engineer who would later become the leading mind behind the success of the first manned Moon exploration, which resulted in the historical landing of Neil Armstrong (1930–2012) and Buzz Aldrin (1930–) to the Moon in 1969. The conflictual framework during which rockets were developed, from the Second World War to the Cold War between URSS and USA, re-emerges in the growing strategic importance of space for military purposes, from reconnaissance satellites to the use of GPS in warfare[21] (cf. Bormann and Sheehan 2009), raising global concerns and calling for demilitarization of space and peace-building (cf. Goh 2004).

Both military as well as civilian applications of space-based technologies lead to another crucial point to reflect upon: their sustainability and environmental impact. Let's offer a few examples, such as the fact that rocket launches damage the global ozone layer (cf. Ross et al. 2009); in recent years, the alternative possibility of developing non-rocket spacelaunches or space elevators has become an emerging subject of research (Bolonkin 2006). Mainstream media such as cell phones and the internet increasingly require the presence of satellites in orbit, which are responsible for the growing amount of space debris, considered a serious threat to space-based activities for possible collisions (NASA Orbital Debris FAQs, n. year). On the other hand, NASA does not fully regard the risk of reentering debris on Earth as an hazard, despite the fact that "during the past 50 years an average of one cataloged piece of debris fell back to Earth each day" (ibid.).[22] According to NASA, the debris surviving the severe heating that occurs during reentry are "most likely to fall into the oceans or other bodies of water or onto sparsely populated regions like the Canadian Tundra, the Australian Outback, or Siberia in the Russian Federation" (ibid.). From a posthuman post-anthropocentric perspective, the fact that the debris may fall in "sparsely populated regions" does not make the problem less urgent, not only because other non-human animals may inhabit those areas, but also because, as NASA warns, "[h]azardous chemicals may be present in debris" (Space Shuttle Columbia and her crew, n. year),[23] thus posing risks of environmental pollution.

In 2008, NASA released an official Statement on the Environmental Impact (PEIS), which takes into consideration the environmental impact of space technology on Earth, but it does not acknowledge its impact on other celestial bodies, such as the Moon or other planets of the Solar System. Critical to this type of anthropocentric and Earth-centric approach, William Kramer underlines: "there is no comprehensive process required…for assessing human impacts on those extraterrestrial environments" (2014, 216). Space technology and space-based human activity shall be analyzed from a view which takes into account their effects

[21]GPS is considered so essential to modern warfare (Gruber and Anderson 2013) that US Secretary of Defense Chuck Hagel in 2013 stated: "To maintain our superiority in space, the Air Force continues to modernize the GPS program" (Secretary of Defense Speech, n. year).

[22]NASA underlines that "[n]o serious injury or significant property damage caused by reentering debris has been confirmed" (NASA Orbital Debris FAQs).

[23]Please note that this warning specifically refers to debris from the Columbia shuttle.

not only on humans and on Earth, but on outer space as well. In order to address this issue, we first need to engage with the question asked by Reinman (2009): is (outer) space an environment? If so, it shall be regulated under specific environmental conditions. In Reinman's opinion, "space at large should not enjoy a moral status equal to Earth" (ibid., 86), as she grants a primacy to Earth based on bio-centric values: "In many ways Earth, with its unique, abundant life, is special. There is nothing quite like it in the Solar System" (ibid.). Although the point raised by Reinman is of key importance to our discussion, from a posthuman perspective, regarding the Earth as "special" because of its life abundance is problematic, being supported by an Earth-centric, bio-centric and quantitative principle which supremacy is not inherently justified; life itself, in fact, is a slippery concept.

The current understanding of life is merely descriptive, not definitive: the border between animate/inanimate is difficult to mark and is often transgressed.[24] Viruses, for instance, exhibit some of the characteristics which are common to organic life, while they are missing others, challenging the biological concept of life itself.[25] More in general, it can be stated that life is not a clearly defined notion; instead, as Michel Foucault noted: "Life…is a category of classification, relative, like all the other categories, to the criteria one adopts" (1966; Engl. Transl. 1970, 161). Going back to Reinman's conclusions, she underlines an aspect of strategic relevance for a posthumanist sensitivity: "humans' actions towards their surroundings will continue to affect people whether we live on Earth or in space" (2009, 86). Let's reflect further upon this point. The non-human agency of matter (Barad 2007), as highlighted within the frame of New Materialism, plays a key role in allowing us to recognize agency to planets, stars and asteroids. The relational onto-epistemological approach of New Materialism makes us think on the possible astro-ecological impacts of Moon mining, or of terraforming in Mars,[26] on the balance of the solar system and, eventually, on their orbits. Even the environmentally-sound concept of space-based solar power (cf. Ernst 2013) should be considered from perspectives others than Earth. Object-Oriented Ontology, and in particular the notion of "Hyperobjects" (Morton 2013), highlights the material viscosity of objects whose performance exceeds both a particular space and a particular time: reading the current opening of the space market from this perspective will unmask the long-term irreversible consequences of our present actions.

Space is the next frontier, where new resources, habitats and life forms are currently being sought: in November 2015, the United States Government passed the "Commercial Space Launch Competitiveness Act "[t]o facilitate a pro-growth environment for the developing commercial space industry by encouraging private

[24] In their comprehensive book "The tree of knowledge: the biological roots of human understanding" (1987), biologists Humberto Maturana and Francisco Varela stated: "Throughout the history of biology many criteria have been proposed. They all have drawbacks" (42).

[25] In his article "Are viruses alive?" (2004), Luis P. Villarreal affirms: "Viruses today are thought of as being in a gray area between living and nonliving" (97).

[26] That is, the hypothesis of changing Mars' climate and surface in order to make it hospitable to humans.

sector investment" (U.S. Commercial Space Launch Competitiveness Act 2015). Although approaching outer space as a resource may spark interest and funding, from an heideggerian perspective, it is ontologically limiting and epistemologically partial, based on an Earth-centered policy sustained by an anthropocentric *Weltanschauung*. Furthermore, the "Space Act" may contravene the international regulations laid down by the "Outer Space Treaty" (1967), a key document ratified by 104 countries, including the US, which still represents the legal framework for space activity. The Office for Outer Space Affairs of the United Nations summarizes the following principles as the main ones sustaining the Treaty:

> the exploration and use of outer space shall be carried out for the benefit and in the interests of all countries and shall be the province of all mankind;
> outer space shall be free for exploration and use by all States;
> outer space is not subject to national appropriation by claim of sovereignty, by means of use or occupation, or by any other means;
> States shall not place nuclear weapons or other weapons of mass destruction in orbit or on celestial bodies or station them in outer space in any other manner;
> the Moon and other celestial bodies shall be used exclusively for peaceful purposes;
> astronauts shall be regarded as the envoys of mankind;
> States shall be responsible for national space activities whether carried out by governmental or non-governmental entities;
> States shall be liable for damage caused by their space objects; and States shall avoid harmful contamination of space and celestial bodies.
> (Treaty on Principles Governing the Activities of States in the Exploration and Use of Outer Space 1967)

As we can see, this document is based on the principle of the common heritage of humankind, according to which "outer space is not subject to national appropriation by claim of sovereignty". Conceived during the Cold War, the Treaty inaugurates a post-nationalistic post-bellic approach to space, setting a new paradigm which has departed from the dualistic imprinting of "us" against "them". Although still within an anthropocentric schemata focussed on the interests of "mankind", the step is huge. For instance, celestial bodies shall be used "for peaceful purposes" and shall not be contaminated; astronauts are considered the "envoys" of humankind.[27] The human frame has been opened and expanded: posthumanism has entered the gates to the heavens.

It is now time to consider the impact of space encountering on human identity and existential insights, by delving into the specific change of perspective brought along by space traveling. This radical shift, known as the overview effect, consists of a series of epiphanies experienced by astronauts looking at the Earth from outer space. In his book *The Overview Effect: Space exploration and human evolution* (1998), Frank White relates such a shift in consciousness to that specific

[27]Following, the scholarly requirements for astronaut pilots should include not only a bachelor's degree in "an appropriate field of engineering, biological science, physical science, or mathematics" (Astronaut Requirements, n. year), but also an education in space ethics, global studies and world citizenship, setting the ground for a new mindset related to space exploration and space migration.

geographical perspective, stating: "Mental processes and views of life cannot be separated from physical location" (3). Humans are embodied beings; their materiality is a process supported and deeply affected by their surroundings. White further asserts this point by emphasizing the fact that the astronauts in Earth orbits and the lunar astronauts have different types of epiphanies: "The orbital astronaut sees the Earth as huge and himself or herself as less significant. The lunar astronaut sees the Earth as small and feels the awesome grandeur of the entire universe...Both programs change the astronaut's perception of the Earth and of his or her own identity, but in quite different ways" (ibid., 36). To White, the overview effect is so significant, that he affirms: "It is possible to grasp the true implications of this evolutionary process only by seeing it from the viewpoint of the universe as a whole, and from that perspective, the Overview Effect may point to humankind's purpose as a species" (ibid., 5). The overview effect is of key importance to space ethics, allowing us to approach the topic of space migration not only from the usual utilitarian perspective, but also from an onto-epistemological standpoint: resonating with Heidegger, space physically becomes "a way of revealing".

10.5 Conclusions

The affects and effects of space travel are life-changing, as Valentina Tereshkova remarks: "As soon as I begin staring into the starry ways in the sky, I physically realize how close they are. Those who have already been in space, yearn with all their hearts and souls to haste there again and again" (2015, 10). Tereshkova recently volunteered for a one-way trip to Mars, believing in a project which, even though not accomplished yet, may soon enough become actual. This chapter responds to the urgency for reflecting on the large-scale ethical implications, socio-political challenges and technological preconditions of space migration. In the first section of this chapter we have demonstrated that in the ancient world astronomical insights had a direct impact on social events, architectonical structures and religious beliefs: knowledge of space was crucial to the understanding of the Earth and to the development of human civilizations. In the second section we have underlined how, in the space race, humans lost their ontological primacy. While humanistic categories such as gender, race, nationality, among others, are still affecting the practices of going to space, the anthropocentric ontological primacy of the human has been challenged. On one side, non humans animals were launched first and have preceded humans in space. On the other side, robots are better suited to survive to outer space conditions.[28] Thirdly, space migration brings to the bioethical debate on human enhancement new terrain of discussion by addressing, among other controversial issues, the search for alien life and the possibility of creating hybrids and chimeras

[28]We can even claim that robots have already migrated to outer space, if we consider that Mars rovers will not be retrieved back to Earth (at least, as far as the close future is concerned).

between human animals and non-human animals, who may be better fitted to live on planets other than Earth, with all the bioethical concerns that crossing such species boundaries may raise. In the third section we have highlighted how outer space cannot be thought separate from Earth: space technology is already causing space debris, an environmental hazard both for spacecrafts as well as for life on Earth. Space pragmatics should be revised by developing sustainable space technology in order to comply with the theoretical premises based on the "Outer Space Treaty" (1967), expanding the beneficial vision of space exploration and space migration, from humans and Earth, to non-human beings and non-human agents, including other planets, stars, natural satellites and asteroids, approaching outer space under specific environmental regulations.

Space exploration and interstellar traveling are setting the conditions for a socio-cultural, bio-technological and geo-political evolution, which is radically challenging the notion of the human, of the cosmos and of life itself. From an onto-epistemological perspective, the narratives of outer space are feeding a posthuman paradigm shift by decentering the Earth from the center of the known universe, and placing hypothetical human and non-human beings on other celestial bodies; furthermore, space migration and the adaptation to extraterrestrial conditions may eventually bring along the evolution of posthumanities. Outer space represents a literal and physical place beyond anthropocentrism, Earth-centrism, biocentrism and life-centrism, although these discriminatory categories are reappearing in human activities and pragmatics in space: this is why space is crucial to Posthumanism as much as Posthumanism is necessary to space. Outer space can finally be seen as the becoming[29] of the human, not only linguistically (as a "post-*humus*", the etymological root of the term "human"), but also ontologically. Outer space has historically performed and continues to manifest as a way of revealing in the processual constitution of human and posthuman identities. Through a comprehensive analysis of past, present and future legacies, this chapter stresses the importance of adopting a posthumanist approach in space migration, in order to manifest, instead of old habits and new wars, desirable futures for humans and non-humans alike.

Acknowledgments Many thanks to Thomas Roby and Ellen Delahunty Roby for proofreading of the chapter.

References

Ackmann, M. (2003). *The Mercury 13: The untold story of thirteen American women and the dream of space flight*. New York: Random House.

Astronaut Requirements. (n. year). NASA website: http://www.nasa.gov/audience/forstudents/postsecondary/features/F_Astronaut_Requirements.html. Accessed January 16, 2016.

---

[29]The term should be read in the deleuzian sense as becoming-outer space (Deleuze et al. 1987).

Ballard, R. A. (2008). Animal/Human hybrids and chimeras: What are they? Why are they being created? And what attempts have been made to regulate them? *Journal of Medicine and Law, 12*, 297–319.
Barad, K. (2003). Posthumanist performativity: Toward an understanding of how matter comes to matter. *Signs: Journal of Women in Culture and Society, 28*(3),801–831.
Barad, K. (2007). *Meeting the universe halfway: Quantum physics and the entanglement of matter and meaning*. Durham et al: Duke University Press.
Bolonkin, A. (2006). *Non-rocket space launch and flight*. Oxford (UK): Elsevier.
Bormann, N., & Sheehan, M. (Eds.). (2009). *Securing outer space*. New York: Routledge.
Bostrom, N., & Roache, R. (2008). Ethical issues in human enhancement. In J. Rysberg, et al. (Eds.), *New waves in applied ethics* (pp. 120–152). Basingstoke (UK): Palgrave Macmillan.
Brady, B. (2013). Images in the heavens: a cultural landscape. *Journal for the Study of Religion, Nature and Culture, 7*(4), 461–484.
Braidotti, R. (2013). *The posthuman*. Cambridge: Polity.
Burgess, C., & Dubbs, C. (2007). *Animals in space from research rockets to the space shuttle*. London: Springer.
Cadbury, D. (2005). *Space race: The epic battle between America and the Soviet Union for dominion of space*. New York: HarperCollins.
Clancey, W. (2012). *Working on Mars voyages of scientific discovery with the Mars exploration rovers*. Cambridge: MIT Press.
Cordrick, H. K. (2008). *Objectivity in the feminist philosophy of science*. London: Continuum.
Cully, A., et al. (2015). Robots that can adapt like animals. *Nature, 521*(7553), 503–507.
Darrin, A., & O'Leary, B. (Eds.). (2009). *Handbook of space engineering, archaeology, and heritage*. Boca Raton (FL): CRC Press.
Deleuze, G., & Guattari, F. (1980/1987). *A thousand plateaus: Capitalism and schizophrenia* (B. Massumi, Trans.). London: Continuum.
Do, S., et al. (2014). An independent assessment of the technical feasibility of the Mars one mission plan—updated analysis. In *Proceedings of the 65th International Astronautical Congress, Toronto*, September 29–October 3, 2014.
Ernst, D. (2013). Beam it down, Scotty: The regulatory framework for space-based solar power. *Review of European Comparative and International Environmental Law, 22*(3), 354–365.
Evans, B. (2009). *Escaping the bonds of Earth: The fifties and the sixties*. Berlin: Springer-Praxis.
Foucault, M. (1966/1970). *The order of things: An archaeology of the human sciences* (A. Sheridan, Trans.). New York: Random House.
Garofoli, J. (2012, July 25). Sally Ride 'never hid,' was just very private. *San Francisco Chronicle* (CA), A1.
Goh, G. M. (2004). Keeping the peace in outer space: a legal framework for the prohibition of the use of force. *Space Policy, 20*(4), 259–278.
Gould, S. J. (1996). *Full house: The spread of excellence from Darwin to Plato*. New York: Three Rivers Press.
Griaule, M. (1965). *Conversations with Ogotemmêli: An introduction to Dogon religious ideas*. London: International African Institute, Oxford University Press.
Gruber, B., & Anderson, J. (2013). Space superiority, down to the nanosecond. *Air and Space Power, 27*(5), 98–119.
Guidetti, R., et al. (2012). What can we learn from the toughest animals of the Earth? Water bears (tardigrades) as multicellular model organisms in order to perform scientific preparations for lunar exploration. *Scientific Preparations for Lunar Exploration, Planetary and Space Science, 74*(1), 97–102.
Hayden, B., & Villeneuve, S. (2011). Astronomy in the upper paleolithic? *Cambridge Archaeological Journal, 21*(3), 331–355.
Hazen, R. M. (2005). *Genesis: The scientific quest for life's origin*. Washington: Joseph Henry Press.
Heidegger, M. (1953/1977). *The question concerning technology and other essays* (W. Lovitt, Trans.). New York: Harper Torchbooks.

Hooton, C. (2014, June 18). SpaceX's Elon Musk: Nasa? 2035? I'll put man on Mars in the next 10 years. *Independent.co.uk*. http://www.independent.co.uk/news/science/spacexs-elon-musk-ill-put-man-on-mars-in-the-next-10-years-9546600.html. Accessed January 16, 2016.

Human Settlement on Mars. (n. year). *Mars-one.com*: http://www.mars-one.com. Accessed January 16, 2016.

Keep, E. (2015, June 2). Mars One reveals true number of applicants. *Matter, medium.com*. https://medium.com/matter/mars-one-reveals-true-number-of-applicants-77b75bf325a3#.4qdr6ahnw. Accessed January 16, 2016.

Kramer, W. (2014). Extraterrestrial environmental impact assessments—A foreseeable prerequisite for wise decisions regarding outer space exploration, research and development. *Space Policy, 30*(4), 215–222.

Langston, S., & Pell, S. J. (2015). What is in a name? Perceived identity, classification, philosophy, and implied duty of the 'astronaut'. *Acta Astronautica, 115*, 185–194.

Launius, R. D., & McCurdy, H. E. (2008). *Robots in space technology, evolution, and interplanetary travel*. Baltimore: Johns Hopkins University Press.

Magli, G. (2009). *Mysteries and discoveries of archaeoastronomy: From Giza to Easter Island*. New York: Springer.

Magli, G. (2013). *Architecture, astronomy and sacred landscape in ancient Egypt*. Cambridge: Cambridge University Press.

Maturana, H. R., & Varela, F. J. (1987). *The tree of knowledge: The biological roots of human understanding*. Boston: Shambhala.

McQuaid, K. (2007). Racism, sexism, and space ventures: Civil rights at NASA in the Nixon era and beyond. In S. J. Dick & R. D. Launius (Eds.), *Societal impact of spaceflight* (pp. 421–449). Washington (DC): National Aeronautics and Space Administration, Office of External Relations, History Division.

Moravec, H. (1988). *Mind children: The future of robot and human intelligence*. Cambridge (MA): Harvard University Press.

Morton, T. (2013). *Hyperobjects: Philosophy and ecology after the end of the world*. Minneapolis: University of Minnesota Press.

NASA. (n.year). *NASA's Journey to Mars*. https://www.nasa.gov/content/nasas-journey-to-mars/. Accessed January 16, 2016.

NASA Orbital Debris FAQs. (n. year). *NASA*. http://orbitaldebris.jsc.nasa.gov/faqs.html#3. Accessed January 16, 2016.

Outer space. (2015). *OED online*. Oxford University Press. http://www.oed.com/. Accessed January 16, 2016.

Over 200,000 apply to first ever recruitment for Mars settlement (n. year). *Mars-one*. http://www.mars-one.com/news/press-releases/over-200000-apply-to-first-ever-recruitment-for-mars-settlement. Accessed January 16, 2016.

PEIS. (2008). *NASA*. http://www.nasa.gov/mission_pages/constellation/main/peis.html. Accessed January 16, 2016.

Petersen, M. B. (2009). *Missiles for the fatherland: Peenemünde, national socialism, and the V-2 missile*. New York: Cambridge University Press.

Quintilian, M. F. [95 AD ca.]. (1920). *The Institutio Oratoria of Quintilian. Book 1* (H. E. Butler, Trans.). Cambridge (MA): Harvard University Press; London: William Heinemann.

Rabkin, E. S. (2002, September 22). The romance of space travel: on the sexual iconography of spacecraft. *Fathom*. http://deepblue.lib.umich.edu/handle/2027.42/55287. Accessed January 16, 2016.

Reiman, S. (2009). Is space an environment? *Space Policy, 25*, 81–87.

Reyes-Ruiz, M., et al. (2012). Dynamics of escaping Earth ejecta and their collision probabilities with different Solar System bodies. *Icarus, 220*(2), 777–786.

Roden, D. (2015). *Posthuman life*. London: Routledge.

Romaniello, G. (2004). *Pensiero e linguaggio: Grammatica universale*. Roma: Sovera.

Ross, M., et al. (2009). Limits on the space launch market related to stratospheric ozone depletion. *Astropolitics, 7*(1), 50–82.

Ruggles, C. (2006). Interpreting solstitial alignments in Late Neolithic Wessex. *Archaeoastronomy, 20*, 1–27.
Russomano, T., et al. (2008). *The effects of hypergravity and microgravity on biomedical experiments*. San Rafael (CA): Morgan & Claypool Publishers.
Said, E. (1978). *Orientalism*. New York: Random House.
Secretary of Defense Speech. (n. year). U.S. Department of Defense website: http://archive.defense.gov/speeches/speech.aspx?speechid=1768. Accessed January 16, 2016.
Shiva, V. (1993). *Monocultures of the mind: Perspectives on biodiversity and biotechnology*. London: Zed Books.
Slater, W. J. (1969). *Lexicon to Pindar*. Berlin: De Gruyter. http://perseus.uchicago.edu/Reference/Slater.html. Accessed January 16, 2016.
Smyth, S. (2013, June 7). First woman to go into space offers to set out on one-way trip to Mars at the age of 76. *Daily Mail.com*. http://www.dailymail.co.uk/sciencetech/article-2337794/Valentina-Tereshkova-First-woman-space-offers-set-way-trip-Mars-age-76.html#ixzz3xLFHmyLv. Accessed January 16, 2016.
Space Shuttle Columbia and her crew. (n. year). *NASA*. http://www.nasa.gov/columbia/help/. Accessed January 16, 2016.
Spivak, G. (1987). *In other worlds: essays in cultural politics*. London: Routledge.
Šprajc, I., et al. (2009). Early Maya astronomy and urban planning at El Mirador, Peten, Guatemala. *Anthropological Notebooks, 15*(3), 79–101.
Tereshkova, V. (2015). *The first lady of space*. Bethesda (MD): Spacehistory101.com Press.
Treaty on Principles Governing the Activities of States in the Exploration and Use of Outer Space, including the Moon and Other Celestial Bodies. (1967). United Nations Office for Outer Space Affairs. http://www.unoosa.org/oosa/en/ourwork/spacelaw/treaties/introouterspacetreaty.html. Accessed January 16, 2016.
Turkina, O. (2014). *Soviet space dogs*. London: FUEL Design & Publishing.
U.S. Commercial Space Launch Competitiveness Act. (2015). *Congress.gov*. https://www.congress.gov/bill/114th-congress/house-bill/2262/text. Accessed January 16, 2016.
Van Allen, J. A. (2004). Is human spaceflight obsolete? *Issues in Science and Technology, 20*(4).
Villarreal, L. P. (2004). Are viruses alive? *Scientific American, 291*(6), 97–102.
White, F. (1998). *The Overview Effect: Space exploration and human evolution*. Reston (VA): American Institute of Aeronautics and Astronautics.
Young, G. (2012). *The Russian cosmists: The esoteric futurism of Nikolai Fedorov and his followers*. New York: Oxford University Press.

Chapter 11
An Urgent Need to Explore Space

Jacques Arnould

11.1 Space Is yet to Be Explored

There was a time when geographers left large areas white on maps, outlining them to indicate territories that had not yet been discovered and explored—areas called *terrae incognitae*, unknown lands. Today, thanks to the daring expeditions of explorers and the tireless satellites that pore over every patch of the Earth, these blank territories have disappeared from maps. Exploration of our planet is more or less complete, excepted perhaps the deep sea bed.

In the late 1950s, space became the new territory to explore. Explorers set off to go where Sputnik had gone, tackled the interplanetary void, and even reached the surface of the Moon. During this time, increasingly sophisticated probes began flying over the planets in the solar system, sometimes even landing on them. For half a century, the successes continued to multiply in spite of the challenges, setbacks and failures.

But today, space exploration seems to be losing steam. Sure, there are plenty of projects for automated missions looking for signs of past or present life on a body in the solar system, or attempting to land on an asteroid in order to bring samples back to Earth. And nations are preparing to send new missions to the Moon, and perhaps one day (in the distant future) to Mars. But the most sceptical (or the most realistic?) among us wonder: how does this compare to the 1960s and 1970s, or the era of the race to the Moon, or the first interplanetary missions? Can one still reasonably call this space exploration?

I would like to defend the idea that not only is space exploration still a very current issue, it is also an urgent need for humanity. I will base my argument on three observations, which are certainties as well as conditions. First, despite the advances and successes of the past sixty years, we still know only a minute fraction

J. Arnould (✉)
Centre Nationale d'Etudes Spatiale, Paris, France
e-mail: jacques.arnould@cnes.fr

J.S.J. Schwartz and T. Milligan (eds.), *The Ethics of Space Exploration*,
Space and Society, DOI 10.1007/978-3-319-39827-3_11

of space: for the most part its 'territories' are still vast *terrae incognitae*. Second, the ability to explore seems to be specific to the human species—an ability that moreover ensures the survival of the human species. Humans explore just as they breathe, but if they stop exploring, they stop breathing! Third, we need to recognise and admit that in order for exploration to occur, there must first be revolution—not just technical or scientific breakthroughs, but also significant philosophical or metaphysical developments—for our basic assumptions about the nature of space accentuate possibilities for its exploration.

11.2 When the Sky Opened up

April 1610. Over the previous winter, Galileo Galilei observed the sky over Padua using a lens he had crafted. He wrote a report of his observations entitled *Sidereus Nuncius* (*The Starry Messenger*), and on 12 March published 500 copies, which quickly sold out. One copy he sent to Johannes Kepler, who offered Galileo his support in an enthusiastic tract entitled *Dissertatio cum Nuncio Sidereo* (*Conversation with the Starry Messenger*), drafted in just eleven days. Kepler wrote: "There will certainly be no lack of human pioneers when we have mastered the art of flight. Who would have guessed that navigation across the vast ocean is less dangerous and quieter than in the narrow, threatening gulfs of the Adriatic, the Baltic, or the British straits? Let us create vessels and sails appropriate for the heavenly ether, and there will be plenty of people unafraid of the empty wastes. In the meantime, we shall prepare, for the brave sky-travellers, maps of the celestial bodies—I shall do it for the Moon and you, Galileo, for Jupiter."[1] In essence, this was the conception of modern astronautics: Kepler was the first to suggest that humans would one day be able to travel into the sky. Until this, Aristotle's cosmology—accepted by the philosophies, theologies and cultures of the West—prohibited such a journey. By virtue of its perfection, order and beauty, the sky was known as the cosmos and considered inaccessible to humans, for the human mind was too encumbered by the flawed and mortal reality of the body. Only souls released from physical constraints through either spiritual exercises or death could hope to cross the celestial threshold and enter this divine sphere. Thus, Kepler's words shattered these taboos along with the crystal spheres on which the planets were thought to move. The sky opened up to human exploration.

In order to establish the foundations of the new science of astronomy, and to defend the strict separation between religious and scientific authorities, Galileo wrote to the Grand Duchess Christine de Lorraine that he had "heard from an ecclesiastical person in a very eminent position, namely that the intention of the Holy Spirit is to teach us how to go to heaven and not how goes heaven." Kepler, however, had already understood that the day would come when the sciences not

[1]Cited in (Koestler 1968, 378).

only explained the movements of the sky, but also made travel into the sky possible. He had already understood that the revolution he was helping initiate, along with Copernicus and Galileo, was not only scientific. Indeed, all of cosmology was being turned upside down. A major shift was occurring in the way humans saw and understood reality—even within the scope of the imaginary and metaphysics.

11.3 When the Imaginary Explores Space

Aristotle's dualist cosmology had such a strong grip on Western thought throughout the centuries that it was impossible to even imagine a human travelling into space. Lucian of Samosata's text *Verae Historiae (True Stories)* written circa 180 A.D. is the exception to the rule, for works of fiction describing a celestial odyssey were virtually non-existent. Lucian warns his readers that he writes of: "things I have neither seen nor experienced nor heard tell of from anybody else; things, what is more, that do not in fact exist and could never exist at all. So my readers must not believe a word I say." He then tells the story of a ship swept up into the air by a violent storm, his journey through the skies for seven days and seven nights, his arrival on a mysterious cosmic island, his meeting with strange creatures, etc. In another work, *Icaromenippus*—the title of which makes clear reference to the myth of Daedalus and Icarus—Lucian writes about the journey of Menippus, a man who has procured wings for himself in order to travel to the Moon and beyond. In the early 14th century, Dante Alighieri invites his readers, guided by Beatrice, to visit the seven spheres of the cosmos; but is this not a sort of dream, a journey of the mind?

It was not until the revolution in astronomy had begun that imaginary tales of humans travelling into outer space appeared. And there were many! In 1638, Francis Godwin published his vision of a wildly enchanting lunar nature and a people more successful than our own, in *The Man in the Moon.* Twenty years later, Savinien de Cyrano de Bergerac published *The Comical History of the States and Empires of the Moon* (1657) followed by *The Comical History of the States and Empires of the Sun* (1662). In 1765, Marie-Anne de Roumier published the seven volumes of *Lord Seton's Voyage Among the Seven Planets,* a true astronomical epic. In 1835, Edgar Allan Poe sent *Hans Pfaall* to the Moon on board a balloon. And I cannot leave out *From the Earth to the Sea* (1865) and *Around the Moon* (1869) by Jules Verne, whose works have marked the minds of multiple generations.

The 19th century was also marked by the end of the systematic exploration of the terrestrial globe led by the West since the end of the 15th century. What was left for humanity to explore, to satisfy its curiosity and thirst for discovery? Space, of course. Space travel was no longer just a theme in stories; it became scientific. More than ever, the astronomical lens had attained the status of a vehicle, thanks to which, sometimes with a good dose of imagination, astronomers peered at the Moon, the planets and the stars. They were even able to produce a map of the surface of Mars that was as accurate as maps of the Earth, or so they claimed. Camille Flammarion, author of *The Plurality of Inhabited Worlds*, dominated the end of the 19th century with numerous publications, in which he asserted that Earth had no special superiority in the solar system and that other worlds were probably also inhabited by life.

There was plenty of imagination in the 20th century, and no lack of authors: from Herbert G. Wells (*The War of the Worlds* in 1898, *The First Men in the Moon* in 1901) to Arthur C. Clarke (*2001: A Space Odyssey* in 1968), the literary genre of space fantasy became very popular. Wells is sometimes considered the inventor of modern science fiction. He was responsible for introducing the public to "scientific wonders" such as the time machine, the element transmuter, and hyperspace. Clarke was a member of the British Interplanetary Society. In 1939, he published an article entitled "We Can Rocket to the Moon—Now!" before developing the concept of geostationary satellites. Meanwhile, cinema had joined literature in the space travel genre beginning with Georges Méliès' *A Trip to the Moon*, which came out in 1902. The films that followed proved worthy successors, including *2001: A Space Odyssey* by Stanley Kubrick, a contemporary of the Apollo programme (1968). In short, the imaginary and the imagination of science fiction and space opera authors seemed as boundless as our universe.

It is important not to underestimate the role and the influence of the imagination and various fields of human culture on the space venture and different nations' space policies. I will take the example of the United States. American historian McCurdy (1997) has shown how the American space programme has been motivated by a romantic dream since day one. What makes his thesis even more interesting is the fact that North American culture does not have the same backdrop that exists in Europe—the literary and artistic landscape I have just described. Lucian of Samosata and Dante, Godwin and Cyrano de Bergerac, from such distant eras, are not known there or at least cannot be said to be of significant influence. In this arena, Americans in the first half of the 20th century were more concerned with observing the sky than travelling into it. Of course, they knew of Jules Verne and his technical fiction, but, McCurdy suggests, they seemed to be wary of it. North America, while not systematically technophobic, seemed to have a hard time fully grasping the technical domain, such that in works of fiction, flight in space was often portrayed as much easier than it is in reality. In this respect, the non-communicative approach of Robert Goddard, the father of American astronautics, was something of a handicap. Yet in spite of all this, things began to change after the Great Depression in the 1930s and after World War II.

It must be said that the success of the Soviets' Sputniks left the American government little choice: faced with this affront to their technological and military supremacy, the United States had to respond quickly, regardless of whether the Soviet threat was real, supposed or imagined. Luckily, the start of the 1950s prepared the public for the launching of ambitious programmes, and got the public interested in space travel. William Ley's writings, Chesley Bonestell's illustrations, Wernher von Braun's conferences, and Walt Disney's attractions all had a significant influence on public opinion. There was also a surge of fascination with flying saucers and encounters with little green men: interplanetary travel and extraterrestrials were the two pillars of the American space dream. While the military branches quickly removed manned flights from their areas of interest, NASA made manned flights one of its primary objectives, encouraged by President John F. Kennedy's speech. Very early on, the dazzling silver spacesuits worn by the seven astronauts of the Mercury programme when they were presented to the press on 9 April 1959 became an iconic image, inextricably tied to imagination surrounding the American space venture.

What can we conclude or deduce from this brief historical panorama, when compared with the current situation? Does the apparent slowing of space exploration—both manned and robotic missions—stem from a lack of imagination, or are our cultures suffering from ideological obstacles similar to those that existed in Aristotle's cosmology, only less apparent? Perhaps humans have finally tired of exploring, of venturing into the unknown, even as scientists explain that the majority of matter and energy making up the universe is still unknown to man; perhaps the origin of this same universe continues to evade us behind the Planck wall… It is time to reconcile with metaphysics.

11.4 The Day Einstein Worried

4 February 1917. Albert Einstein was preparing to address the members of the venerable Royal Prussian Academy of Sciences, founded in Berlin in 1700. To his friend and colleague Paul Ehrenfest, Einstein wrote, "I have committed something in the theory of gravitation that threatens to get me interned in a lunatic asylum" (1993, 91). What a surprising statement, coming from Einstein! Have we ever known a researcher to be committed to an asylum for coming up with a new hypothesis, a new theory? Yet Einstein's fear was no exaggeration: it is similar to Charles Darwin's fear when he told John Hooker that, contrary to the commonly accepted and well-regarded doctrine, living species are not unchanging. "It's like confessing to a murder," the British naturalist wrote to his colleague in a letter dated 11 January 1844, fifteen years before the publication of *The Origin of Species*. These two heavyweights of science sensed beyond what they could measure the impact their respective work would have on the sciences they studied and, more broadly, on the way in which reality (whether physical or biological) was perceived and understood. They knew that the institutions dedicated to the quest for and dissemination of knowledge were not exempt from dogmatic reflexes and ideological dictatorships; they knew that these worlds can include the descendants of the ecclesiastical inquisitors, capable of reducing overly innovative or revolutionary minds to silence. They also knew that their own theories—even if they proved accurate and were accepted by their colleagues, and even if they brought about scientific revolutions in their respective disciplines—would all the same one day or another be questioned, challenged and made obsolete by new findings and new theories. Scientific research is an incredible edifice in which the bold risks of

yesterday serve as a scaffolding for the constructions of today, until they in turn are partially destroyed and buried, serving as the foundation for future progress. How could a scientist, even one of Einstein's stature, not be seized by fear when the time had come to expose his research to his peers? What made Einstein's fear all the more justified was that his talk on this occasion, on 8 February 1917, was entitled *Cosmological considerations on the general theory of relativity.*

The term 'cosmological' is clue enough as to the reason for the concern Einstein mentioned to his friend Ehrenfest. He already knew that what he was about to say would make a mockery of Kant's assertion that scientists must not address the matter of cosmology and not speak of the Whole; and Kant was not the only one to think that this subject should be left to philosophers and theologians. Einstein had dared to break this rule. What he suggested to his fellow scientists was to represent the universe in the form of a cylinder around which particles' trajectories were wound; in this first model of modern cosmology, matter and energy were evenly distributed, presenting no unique traits except local irregularities, which were without design and had no cosmic significance. Lastly, Einstein's universe was at rest.

Einstein's talk in Berlin in the winter of 1917 opened a new chapter in the history of the sciences and the exploration of reality: from this point on, it was a question of the theory of relativity and quantum mechanics, the Big Bang theory and the multiverse.

While it is not difficult to acknowledge and even evaluate the scientific revolution thus based on cosmic and physical reality, one must not ignore the philosophical and metaphysical consequences of this same revolution. If some philosophers have taken up this theme (Henri Bergson was one of the first to take an interest in it; philosophers of science also), a number of scientists have understood their work's impact in terms of human thought. Einstein, Niels Bohr, Paul Dirac, Louis de Broglie and Werner Heisenberg are among those who have undertaken this metaphysical approach with courage, honesty and lucidity, from a perspective free of any sectarianism, in a true spirit of research and exploration.[2]

11.5 Metaphysics, a Condition for Exploration

The collection of Aristotle's writings compiled by Andronicus of Rhodes circa 60 B.C. separated the books on nature, *phusike akroasis* (Physics), from those that came later, *meta ta physika* (Metaphysics). In other words, the word 'metaphysics' originally had only an 'editorial' signification: it qualified the writings by Aristotle that were published after his writings on physics. However, Platonic thinkers extended this meaning to include the philosophical approach itself: thus, metaphysics encompassed the discipline and ideas regarding realities beyond physics.

[2]See (Arnould 2015).

As a discipline it deals with the fundamental questions such as the immortality of the soul, the question of God, the reasons for the existence of evil and the meaning of life, as well as the study of "the being as a being." i.e., ontology, the study of substance. It addresses all of the questions, it would come to be agreed, that do not as such fall within the fields of the physical sciences, any more than they fall within the scope of space exploration. They may, however, share the same borders!

Indeed, it would be difficult to ignore the fact that (as illustrated by the experience of the scientists I have just mentioned) the search for the physical causes of what is real inevitably leads to reflections on its origins—its original principles—and therefore ends up dealing with metaphysics. On this subject, Werner Heisenberg reports a conversation between scholars in October 1927 in Brussels. Paul Dirac launched into a long diatribe against religions, which he described as nothing more than "a jumble of false assertions, with no basis in reality." Wolfgang Pauli at first remained silent, but when asked to give his opinion said, "Well, our friend Dirac, too, has a religion, and its guiding principle is *There is no God and Dirac is His prophet*," which set everyone laughing, including Dirac himself.[3]

There are plenty of anecdotes (for anecdotes is all they are) such as this in space exploration, anecdotes where the power of the "rational" does not keep the "irrational" from emerging. While today it would seem reasonable to attribute to Nikita Khrushchev rather than Yuri Gagarin himself the note in which the first Soviet astronaut did not see God during his first flight into space, there have been many allusions to religion and many religious statements of all kinds have been made during space missions. A short time after the success of the Apollo 11 mission, the theologian François Russo wrote very astutely: "While the space conquest does not fulfil the infinite need which is in man, it does provide precious analogies. *Space*, said Baudelaire, *is the place of symbols*. In this sense, the man's conquest of space and landing on the moon will not establish us in the illusion, and will not turn us away from this more obscure and interior infinity which alone can fulfil our truest and deepest desire. Far more than that, it will help take us there. Was this not what could be read in Borman's prayer for the Apollo 8 flight last Christmas?" (Russo 1969, 175). Because it concentrates the universe into human dimensions, and because it expands humanity to the dimensions of the universe, and because it is excessive by nature and reaches towards a horizon which will always be inaccessible, the space venture conveys and is itself a profoundly human experience, endowed with a sacred dimension which leaves no one completely indifferent.

Has our own era sufficiently recognised and accepted the metaphysical dimension as a condition for exploration? We can have doubts about this, and we must at least take this question very seriously, without confusing metaphysics and religion or symbolism and ritualism, and consider the matter of the sacred.

---

[3]See (Heisenberg 1971, 85–87).

11.6 The Sacred, Reconsidered

Has space not been declared sacred, as the sky once was, perhaps even since before the sky was thought of this way? Its incommensurable elevation and its terrifying infinity, its cold immutability and its formidable power imposed themselves on the newly-budding human consciousness, in the experience of a primordial separation and crushing dialectic.

Indeed, what is set aside and separated, invested from an intangible value and an inviolable character is declared sacred. The *fanum* area of a temple is separated from the *profanum* area that surrounds it. To enter the sacred area, worshipers must remove their shoes, wear special clothing, wear or remove a head covering, purify themselves with water and/or trace particular signs on their bodies. Sometimes a place within the temple has an even holier nature, such as the choir or tabernacle in churches, the minbar in mosques, the garbhagriha in Hindu temples, and the Holy of Holies in the Temple of Jerusalem. Whatever its name, this area is off limits to common worshippers or lay people; only priests are allowed to come near it, touch it and to enter it, only after completing additional rituals, failing which they can be accused of profanation. Anything that is used to touch the divine or which has touched the divine (in order to serve something divine or through a special action) is also deemed sacred. In many religions, a mother who has just given birth is declared sacred and even consecrated because she has brushed against the origin of human life—the very mystery of life itself. She cannot have normal contact or ordinary exchange with other beings, even those who are the most familiar, for fear that they will be soiled. In other words, there can be no mixing of the sacred and the profane. Like a vase that has been used in a religious rite and has thus also touched the sphere of the divine, a new mother must be stripped, purified, cleared of any trace of the divine before she can return to the community of ordinary lay people.

Also, one who has committed a crime against the gods or the State is also declared sacred. He is shut away, separated from the religious or social group. Generally, only his death can restore order and peace to the world.

Finally, anything on which a society, tradition or culture places special worth is also declared sacred. Thus, life is considered sacred, with special attention to its first and last instants. It is not out of reach or untouchable: a soldier can deliver death, but is also exposed to it; societies can practice the death penalty; and a doctor can perform surgical acts on a patient, which would be illegal if performed by anyone else. Nevertheless, all of these acts remain acts of transgression, and they come with risks.

Thus, it is not surprising that space was, perhaps the first, declared sacred. It both frightens and captivates the human mind, possessing the traits *tremendous* and *fascinating* which the German philosopher Rudolf Otto distinguished. With the beginning of its exploration and conquest, space did not entirely lose its sacredness: astronauts, cosmonauts and taikonauts themselves were, and still are, objects of the process of sacralisation, albeit involuntary. From the colour of their clothes (the white of purity, the red of sacrifice) to the ascetic rigour of their training, the strict selection process and the aura of their fame—everything was in place to make them into heroes—sacred figures. Made sacred again when they returned from space, the American astronauts plunged into the waters of the Pacific Ocean, as if in a bath of purification, and faced their final trial: quarantine. The sky, which had become space, had not lost any of its power; its exploration had even elevated a small portion of humanity to sacredness!

Humans need the sacred, not as much to adore it from afar or to be dominated by it (the darker side of the sacred), but on the contrary, to be drawn to it, to be called by it to give the best of themselves in order to come closer to it and even to reach it and be able to touch it. The very nature of the sacred allows it to be transgressed! This idea may be surprising but is none the less true: the process of making a reality sacred necessarily involves specifying the conditions for its transgression, whether by adhering to an era or performing a rite, etc. At the centre of the temple of Jerusalem, the Holy of Holies was the Jewish faith's sacred place *par excellence*; a place no one could enter—except the High Priest, once a year, to pronounce the name of God which otherwise could not be uttered. And so it is for all Holiest of Holies, for all places established or recognised as sacred by humanity: they must be able to be transgressed to have a true influence over those who respect them. Thus, the sacred is not (or should not be) a cause for prohibition, frustration or castration, but rather a source of liberation, invention and inspiration. There is universal truth to the two-thousand-year-old teaching of the Rabbi of Palestine, "The Sabbath was made for man, not man for the Sabbath"; in other words, the sacred is made for man, not man for the sacred.

11.7 Avoiding the Pitfalls of Dogmatism

Failing to heed these wise words has unfortunately led to unreasonable and excessive behaviour at various periods in human history. Ritualism is limited to religious practices, but dogmatism is the most dangerous ideological and social form of ritualism. It is important to first establish a relevant definition of dogma, and to remember that dogmas are used by not only theologians but also philosophers and scientists.

Dogmas are in essence reference points, such as theories or worldviews, which allow researchers to diverge from the lighted paths, to step outside the bounds of established and confirmed knowledge, in order to venture into dark seas and lands of the unknown, the *terrae incognitae* of explorers, without losing sight of familiar territory and the milestones of knowledge. All scientific research, all intellectual or geographical exploration is necessarily based on a pre-existing body of dogmas, both known and unknown. It is up to the researcher or explorer to be capable of taking distance from the dogma, veering away from it and stepping out of bounds in order to trigger the type of breakthroughs that Kuhn (1962) qualified as revolutions.

The dogmatic attitude is to refuse to accept the slightest distance from dogmas or the slightest transgression of boundaries. The theologian Karl Rahner likens dogmatism to the behaviour of a drunk who has lost his keys; in the dark of night, he latches onto a lamppost to prop himself up, even though this limits the area he can search to the circle of light shed by the lamp. It is an amusing image, but accurate none the less: dogmatism is one of the most powerful obstacles to spiritual vitality, as well as to research and scientific development.

There are many reasons for adopting a dogmatic attitude, but the main drive is certainly fear—fear of the unknown, fear of change, fear of danger, fear of struggle. Kuhn is right when he says that a scientific approach requires overcoming one's fears: "The operations and measurements that a scientist undertakes in the laboratory are not '*the given*' of experience but rather '*the collected with difficulty*'" (Kuhn 1962, 126; emphasis added).

Another reason for dogmatism is a preference for wonder over surprise. Wonder is not reserved solely for artists and aesthetes; it is the basis of many writings in natural theology and spirituality, which evoke the beauty to behold in the spectacles of the sky and the earth. But wonder can also be found in scientific works. In his *Etudes de la nature* (*Studies of Nature*), Bernardin de Saint-Pierre slips easily into this sentiment, shamelessly weaving in the analogical approach and a teleological perspective, running the risk of admitting as self-evident, that which remains to be proven! One of the most famous examples by this French author, who died in 1814, is the existence of a superior intelligence who decided that melons would have ribs in order to make them easier to share, which is presented as "proof" of the existence of God, the "great clockmaker" ("*grand horloger*"), as in William Paley's *Natural Theology*, published in 1802.[4] To what extent is the discourse of the anthropic cosmological principle overly imbued with a tone of wonder, to the detriment of scientific reason? In any case, wonder comes with the risk of forming or even imposing a worldview that is too particular to allow for the sudden emergence of something different, the unknown, amidst the body of certainties and knowledge.

Thus, it would seem that when it comes to creating and promoting research and exploration, surprise is preferable to wonder. Surprise, better than wonder, avoids the pitfalls of dogmatism and blind devotion, and it resists against the trivialisation that stifles the imagination. Philosophers like to name wonder as one of the starting points for their positions and labours. In *Metaphysics*, Aristotle writes: "For it is owing to their wonder that men both now begin and at first began to philosophize; they wondered originally at the obvious difficulties, then advanced little by little and stated difficulties about the greater matters, e.g. about the phenomena of the Moon and those of the Sun and of the Stars, and about the genesis of the Universe. And a man who is puzzled and wonders thinks himself ignorant."[5] And Aristotle was right. Without neglecting to regard the reality of the world with wonder, we must continue to be surprised by what our scientific instruments reveal about it on an evolving scale. We need this wisdom.

---

[4]See de Saint-Pierre (1784), Paley (2006).

[5]Aristotle. *Metaphysics*. Book A, I, 1.

11.8 The Urgent Need to Explore Space

Today, more than half a century after Sputnik and Gagarin, Armstrong and Aldrin, there are plenty who predict or call for the end of space exploration, not including those who fall under a perspective of dogmatism or ideology. The validity of certain arguments in support of this announced or expected end needs to be considered: e.g., the limits of our human means, both technical and financial, compared to the immensity of the universe; the multiple and in some cases unavoidable dangers that humans face in space; the lack of immediate usefulness of scientific programmes relating to space; and the urgent need to focus our efforts on protecting the future of Earth, etc. Conceding to this reasoning would be a grave error.

It is of course out of the question to ignore the dangers facing our planet and its inhabitants. In fact, space techniques have become an essential tool for monitoring climate change and the evolution of the Earth's oceans and vegetation. But it would be as dangerous to end space exploration as to declare forecasting and modelling projects futile. We need a vision of the future to set the course for our odyssey, and only the future can give meaning to the past. In the same way, space exploration—while stimulating the human imagination and demanding a certain force of character—helps combat various dogmatic tendencies by requiring us to undertake frequent revolutions.[6]

Nothing can be taken for granted, especially since the technological challenges that lie ahead will require us to continually reassess our vision of reality. Increasingly, the tales of the explorers of old are being replaced by pictures taken by robots in the farthest corners of our visible universe, such that Earth can now be received right at home. In the words of the French sociologist Régis Debray: "The shift from globes to the appliances department in big-box stores (the audiovisual section), planet Earth has been both miniaturised and domesticated. It can be delivered to our homes like a refrigerator or vacuum cleaner" (1992, 412). Thus, we need to learn how to discover these images (and sometimes sounds) without letting the technical prowess and aesthetic wonder overpower or stifle our sense of surprise—the very spirit of exploration—and be willing to embrace and undertake the revolutions that will lead to the discovery and knowledge of territories heretofore unknown.

References

Arnould, J. (2015). *Sous le voile du cosmos*. Paris: Albin Michel.
de Saint-Pierre, J. H. B. (1784). Études de la nature, chapter XI, sec. *Harmonies végétales des plantes avec l'homme*.

[6]See (Munevar 1998).

Debray, R. (1992). *Vie et mort de l'image. Une histoire du regard en Occident*, vol. 412. Paris: Gallimard.

Einstein, A. (1993). *Œuvres choisies, tome 3*. Paris: Seuil.

Heisenberg, W. (1971). *Physics and beyond. encounters and conversations*. New York: Harper & Row.

Koestler, A. (1968). *The sleepwalkers: A history of man's changing vision of the universe*. London: Pelican.

Kuhn, T. (1962). *The structure of scientific revolutions*. Chicago: University of Chicago Press.

McCurdy, H. E. (1997). *Space and the American imagination*. Washington & London: Smithsonian Institution Press.

Munevar, G. (1998). A philosopher looks at space exploration. In G. Munevar (Ed.), *Evolution and the naked truth* (pp. 169–79). Ashgate.

Paley, W. (2006). *Natural theology*. Oxford: Oxford University Press.

Russo, F. (1969). L'opération apollo. *Études, 331*, 163–175.

Part IV
Planetary Protection and Microbial Value

Chapter 12
The Ethical Status of Microbial Life on Earth and Elsewhere: In Defence of Intrinsic Value

Charles S. Cockell

12.1 Introduction

The most likely type of life that we will encounter during planetary exploration, if we encounter it at all, will be microscopic life. On Earth, organisms that are circumscribed by this description include single-celled bacteria, archaea and eukaryotes as well as some microscopic multicellular organisms such as nematodes.

Although I do not specifically exclude multicellular life, this essay is principally directed towards those organisms that dominated planet Earth for at least 2 billion years before the rise of atmospheric oxygen: single-celled microscopic organisms (Falkowski 2015). Complex multicellular animals have been a pervasive part of the Earth's biosphere for approximately 600 million years and so, at least in terms of the history of terrestrial life so far, single-celled microbial life has dominated the history of life on Earth. Even today, we live on a planet with $\sim 10^{30}$ microorganisms (Whitman et al. 1998) whose diversity and quantity of genetic information is enormous. The majority of the DNA of the planet is to be found in microorganisms (Landenmark et al. 2015).

Given the link between complex multicellular life and the rise of oxygen, it would appear that such life, including intelligence, requires quite specific changes in planetary atmospheric conditions (Catling et al. 2005). Complex multicellular life follows, evolutionarily, single-celled microscopic organisms. Therefore, it is reasonable to suppose that elsewhere we would expect more simple microscopic organisms to precede complex multicellular life and to be present either as a component of extraterrestrial biospheres or to be the only component.

It is therefore a compelling priority in extraterrestrial ethics to define an ethic for microscopic organisms. Developing an ethic for extraterrestrial microscopic

C.S. Cockell (✉)
School of Physics and Astronomy, James Clerk Maxwell Building, Edinburgh EH10 4EP, UK
e-mail: c.s.cockell@ed.ac.uk

J.S.J. Schwartz and T. Milligan (eds.), *The Ethics of Space Exploration*, Space and Society, DOI 10.1007/978-3-319-39827-3_12

organisms at the current time has a number of possible uses: (1) it could help advance our discussion on the ethics of extraterrestrial life even prior to its discovery so that we are prepared should we ever find it; and (2) regardless of the discovery or not of extraterrestrial life, debating the ethics we apply to hypothetical extraterrestrial microscopic organisms provides a basis for considering our ethical approach to microscopic organisms on Earth. As they represent the majority of the biomass and diversity of life on Earth, establishing an environmental ethic for microbial life is important and astrobiology provides a broad context and external perspective to debate the value of microorganisms in general.

12.2 Instrumental and Intrinsic Value

Although not ubiquitous, a common approach to ethics is to recognise two types of value in objects. The first is instrumental value, a statement about how useful an object is to people. Microbes have enormous instrumental value to humans in the production of drugs, food, alcoholic drinks and a variety of other chemical transformations that use the metabolic, including the fermentative, capabilities of organisms (Cockell 2005a). Terrestrial microorganisms not only have this instrumental value but one could also propose that they have a special subset: survival value (Cockell 2005a). As they are involved in cycling the major elements, such as sulfur, iron and carbon, through the biosphere and they carry out vital transformations such as the fixation of atmospheric nitrogen into biologically available nitrites and nitrates, they are vital to the survival of the rest of the biosphere.

Extraterrestrial microscopic organisms, if they exist, would likely have instrumental value as objects of scientific study, particularly if they presented an entirely novel biochemistry. We would collect them to study the mechanisms of information storage (their equivalent of DNA), the chemical transformations they undertook and the types of uses they might even have in industry. These organisms would not be instrumentally valuable as part of the Earth's biosphere as microorganisms are on Earth, but their value would lie in the insights they would give into the process of evolution and how it may operate, and the products it can give rise to, elsewhere. From an anthropocentric point of view, they would clearly have instrumental value, which is one of the least controversial types of value that we can ascribe to terrestrial and extraterrestrial life.

A second type of value sometimes ascribed to objects is intrinsic value. Intrinsic value is a more controversial concept that proposes that objects in the natural world have some claim to more consideration independent of their use to people, namely that they have a value inherent in their existence.

The question of value can be considered either as something intrinsic to the object, independent of any valuer, or as something that emanates from a valuer (in which case it might better be described as extrinsic value, but intrinsic value has been a traditionally used term).

From a reductionist point of view, it is difficult to see how anything could have value, whatever the definition of this term, without something to observe it. Without such an observer, all objects in the Universe are simply matter that may or may not be carrying out chemical reactions, and some of those chemical reactions may or may not lead to self-replication of that matter or evolution as we understand life to do. Therefore, I will assert here that whether we are talking about instrumental or intrinsic value, these values require a valuer to see the object. This view would be consistent with Callicott's notion of 'truncated intrinsic value' (Callicott 1986) which suggests that life can be valued for its own intrinsic value, aside from its instrumental needs, but that this value is still projected on to the object by people. This view is in contrast to Rolston (1988), who defended a notion of intrinsic value that is inherent in the object, independent of any valuer. It is difficult to see how this can be defended without some type of physical explanation of exactly what this intrinsic value resides in and precisely why something in the Universe would have intrinsic value without any valuers. Regardless, the argument may not be important anyway because whether intrinsic value is something inherent in an object or something projected on to it, we definitely do need a valuer for that value to become of any ethical relevance. If not, the object is merely an object in the Universe unmolested by any intelligence going about its business and therefore not subject to the conclusions of any ethical codes. This is an underlying assumption I adopt here.

Throughout the history of ethics, the major debate has centred on where we draw the line of moral relevance, or in value terms, where we draw the line between things that have intrinsic value and those that do not. To ratio-centric ethicists such as Kant (2004), the line of moral consideration exists between things that can reason and things that cannot, placing humans within the bounds of having intrinsic value and other organisms outside.

Some ethicists, such as Bentham (1823), Regan (1983) and Singer (1975), draw the line at sentiency, or the ability to feel pleasure and pain. For such ethicists, dogs fall within the purview of objects that have intrinsic value but trees do not.

For other ethicists, such as Taylor (1989) and Goodpaster (1978), simply being alive is sufficient for having intrinsic value.

An assumption underpinning all these lines of thought, which becomes particularly relevant in the extraterrestrial case, is that we know how to define whether something is alive or not in the first place. The definition of life is particularly important for those ethicists who assert that being alive is sufficient for moral consideration. Since we lack an agreed coherent definition of life—and it may be a non-natural kind, a mere human semantic definition (Cleland and Chyba 2002; Ruiz-Mirazo et al. 2004) rather than some physical set of characteristics that we can exactly define—then a clear-cut separation between life and non-life may be impossible. We might use a working definition, such as that attributed to NASA Scientist Gerald Joyce that life is a "self-sustaining chemical system capable of Darwinian evolution" (Benner 2010). However, the lack of a clear and agreed definition cautions us to maintain an open mind and not to exclude objects from the sphere of ethical debate because they do not fit within a neat human definition of life.

Nevertheless, for any ethicist who claims that simply being alive is sufficient for moral consideration, it remains necessary to define where that line is drawn, even if ultimately it is drawn using an expedient operational definition of life.

12.3 A Microbial Ethics Point of View

In previous papers, I have attempted to defend a view of the microbial world that includes an intrinsic value argument (Cockell 2004, 2005a, b, c, 2008, 2010), namely that microbes should be afforded a moral significance beyond purely their instrumental value to humans, and I have discussed the implications of such an ethic for extraterrestrial life.

The argument that microbes have intrinsic value could be based on their possession of rudimentary interests. We know what is good or bad for a microbe based on physiological attributes, although of course a microbe does not know it is being treated badly. A pertinent question is then to ask what makes microbes different from machines. We know what is good or bad for a tractor, but we do not claim that it has intrinsic value. What separates a microbe from a machine is that microbes have latent tendencies and evolutionary capacities that might demand from us an appreciation of a value in them that transcends their use as resources. If a community of microbes on a planet has the potential to diversify or even to eventually develop into a biosphere containing complex life, these potentialities are frustrated by the destruction of those organisms.

Based on their possession of rudimentary interests, we might argue that individual microscopic organisms have some claim to moral consideration and relevance. However, we cannot live our lives without destroying microbes when we clean our houses and generally carry on our everyday activities. Therefore, such an ethical view is often, but not always, impractical. There are often situations when we can preserve microorganisms. We do not have to wantonly destroy microbes and the communities in which they reside, in a lake for instance, to build a new housing estate. If we think that microbial communities have intrinsic value we could preserve part of the lake or seek to build around it. However, considering intrinsic value for individual microbes is clearly practically difficult in most cases.

Persson (2012, 980) said of this view that "Cockell tries to handle this problem by saying that his ethics can therefore just be a principle that cannot be implemented" and goes on to observe that ethics must be prescriptive and that if an ethical framework cannot be implemented, then it cannot be an ethic at all. However, previously I have posited (Cockell 2005a, 385) that "many individual microbes can be protected when it is possible" and go on to provide an example of a well-ordered microbial community growing around the edge of a lake which we might walk around rather than through, thus disrupting or destroying it. This view is similar to Attfield's views on trees (Attfield 1981). He defended the intrinsic value of trees but recognised that there are situations when we need to cut them down. He stated: "There are, of course, in practice, ample grounds for disregarding the interests of trees at most junctures. But this is not to make trees of no ethical relevance in themselves" (ibid., 52). Similarly, we may be forced to disregard the interests of microbes in many situations (but not all), but this is not to make individual microbes of no ethical relevance in themselves. The inability to implement an ethic at all junctures does not render it "no ethic at all."

However, can we construct an ethic that manifests itself on larger scales—the scales of microbial ecosystems? I have previously argued for a type of 'biorespect' for microbial life. A biorespect encompasses a respect for individual microbes through to communities. On what basis is such a 'respect' constructed? I have suggested that: "Part of our reverence for the microbial world must surely reside in the awe we feel for the sheer scale of their biogeochemical processes and their longevity on Earth. Microbes have mastered and influenced the surface of the Earth in profound ways. How is it not possible for us to show respect for such organisms" (Cockell 2005a, b, c).

Such views have no basis in any objective quality or feature of microbes. From a scientific point of view we should desist from attempting to understand empirically what this sort of statement really *means* from a biological point of view. However, it is a statement rooted in the idea that as latecomers to the evolutionary story of Earth, we should project on to the microbial world a sense of reverence and importance beyond purely their instrumental use to us. It is a form of intrinsic value that recognises the non-instrumental value of microbes. This view of microbes then cautions us to behave in a way towards the microbial world that is more than an assessment of their instrumental uses. I will return to this later.

The problem with terminology such as 'biorespect' is that 'bio' is implicitly a reference to terrestrial life. Although we might expect extraterrestrial life to have similar characteristics, at least in terms of growth, reproduction and evolution, as terrestrial life, we need a term that more successfully encompasses any type of life form. Another term could be 'telorespect' or 'teloempathy', which is derived from the Greek *telos* or purpose (Cockell 2010). The use of telos in this context does *not* imply some pre-defined purpose or goal-oriented nature of life (evolution has no goal), but rather the characteristic that life has of growing, reproducing and evolving in accordance with instructions laid down in its information storage system—the characteristics of living things that bring it within the realms of ethical debate in the first place. Telorespect or teloempathy merely captures our recognition that extraterrestrial life, including life independently evolved from the biology that we know on Earth, places demands on our behaviour if we think it has intrinsic value.

12.4 The Demands of Intrinsic Value for Microbes

Deciding that microbes have intrinsic value does not make the ethical challenge easy. Similarly, suggesting intrinsic value for trees makes decision-making difficult. But ethical frameworks do not need to make things simple and nor do they necessarily have to provide a completely clear-cut framework for decisions at all junctures, although obviously it is helpful if they do. They might instead mollify the worst excesses of a view of the world that is entirely rooted in instrumental values and warn us at all junctures to consider carefully competing interests, even if those competing interests cannot always be objectively quantified and assessed.

Smith (2009) criticised intrinsic value. One of his observations is that intrinsic value, if applied to a wide range of entities, makes it increasingly difficult to make ethical decisions since "instrumental tradeoffs are not allowed," and that it "serves as a trump card against any attempt to argue that we should ignore the welfare of such an entity in favour of some instrumental use." Similarly to Varner (2002) and Sturba (1995), I have taken the implicit view that there are degrees of intrinsic value. These theories, for example that of Varner, see differences in intrinsic value rooted in whether organisms have such things as 'biological interests', which would apply to microorganisms, and 'non-categorical desires' and 'ground projects', whereby the latter do not apply to microorganisms but may apply to more complex sentient or reasoning organisms.

Smith (2009, 270) asserted that there is a problem with this view in that: "such a move robs the intrinsic value account of its original motivation. The whole point of distinguishing intrinsic from instrumental value in the first place was to prevent any kind of haggling over who is more important than whom in different circumstances." However, an alternative view of intrinsic value is not to see it merely as a way of trumping instrumental value, but that it asserts a value other than mere use to humans. This value may not be an either/or decision, but may well come in gradations depending on the characteristics of given entities. The intrinsic value of a human who has ground projects and conscious plans ahead of them is greater than the intrinsic value of microbes, who merely have basic biological interests. Does such a position make ethical decisions easier? Almost certainly not. What it does do is inject a dimension of moral consideration into ethical debate that considers life to have more than its use as a resource. Smith (2009, 274), in defending ratio-centrism as the basis for an ethical division, stated that "what makes ratio-centrism so powerful is that it not only provides a principle of division, it also establishes a common system of measuring moral worth and thus a means for adjudicating ethical disputes." Ratio-centrism certainly does make life easier. If the ethical franchise is narrowed, decisions become successively easier as more entities are excluded. However, the success of this approach hardly provides an argument for its use. It might even invite ad hoc divisions motivated by a desire to provide simplicity and clear-cut methods of ethical adjudication.

Applying intrinsic value to entities other than people and in the context here, to microbes, might have a variety of consequences. In a previous paper, I suggested three possible consequences of thinking that microbial life has intrinsic value (Cockell 2010, 86):

1. No duty to preserve individual microscopic organisms when such individuals put constraints on human activities that are considered to be part of daily life, but a duty to preserve individual organisms when we reasonably can. This ethic

implies a hierarchy whereby human instrumental needs must sometimes outweigh microbial intrinsic value.

2. A duty to implement environmental policies that protect ecosystems and communities of microscopic organisms. The jump from individuals to collectives (i.e. ecosystems) is not an ethically simple one, but I make the assumption here that because microorganisms are part of communities or ecosystems we cannot actually conserve individual microorganisms without protecting the ecosystems to which they belong, so that an ethic directed towards individual microorganisms must necessarily also be directed at the larger scale of organisms.
3. A general duty to show respect towards microscopic organisms in our activities.

I recognised that these statements do not provide clear-cut guidelines for adjudication of interests and suggested that we should be concerned when:

1. We begin to systematically destroy communities of microscopic organisms of which individual microorganisms are a part; the localised killing of a community near a station on a planetary surface would be acceptable, provided that we did not systematically, for whatever reason, deliberately destroy these communities on a planetary scale.
2. We cause the destruction of the biological diversity of microscopic organisms on other planets by systematic environmental change or destruction.

Again, these sorts of statements are vague but to reiterate, they caution us to think about our impact on life and to consider whether we could do things in a way that reduce or eliminate wanton destruction of life. Clearly, these guidelines are not failsafe. What would happen if a single microbe released from a planetary station compromised an entire planetary biosphere? In such circumstances, some supposedly allowable local destruction could have planetary-scale implications that were unforeseen. Avoiding such situations is partly a matter for ethics, but also very much a matter for microbiology. We would be wise, if we find out there is life on another planet, to greatly expand our scientific understanding of that life and its weaknesses and vulnerabilities before making ethical adjudications about whose interests are a priority. Is it likely that the introduction of a single non-indigenous species of microorganism could result in a widespread collapse or serious dysfunction of an indigenous biosphere? This is a question for microbiologists to answer. Positing intrinsic value of microbes based on a sense of respect for it encourages us to engage in the scientific investigation of extraterrestrial life so that we are in a position to attempt to adjudicate the potential impact of our actions and to decide how best to prevent its destruction.

One must also recognise that proposing that microbes have only instrumental value might make ethical decision-making easier, but it does not eliminate uncertainty. The introduction of terrestrial microbial life into an extraterrestrial environment could compromise the instrumental use of that life as a scientific resource. Deciding whether robotic or human exploration will compromise that scientific value is no simple task and not necessarily made easier by rejecting the intrinsic value argument.

The views elaborated on in this paper are similar with some other writers, who, if not explicitly, make statements that imply intrinsic value for extraterrestrial microbial life. This does validate the intrinsic value argument, but it suggests some independent common lines of thought. Sagan (1980) stated that if life is found on Mars, "Mars then belongs to the Martians," although the practical implications of such a statement are uncertain. McKay (1990) suggested that if Martian life was found "it has the right to continue its existence, even if its extinction would benefit the biota of Earth." Lupisella (1997, 1999) urged the view that "first and foremost is a requirement that the rights of any indigenous biota be respected," and Race and Randolph (2002), in referring to life on Mars, suggested that the organisms should be allowed to follow their "natural evolutionary or development trajectory."

Few of these authors elaborate on the ethical foundations of these statements, namely exactly *why* we should hold these views. However, it is probably not wide of the mark to hazard a guess that buried within these statements is the sense that it shows a disrespect to destroy extraterrestrial life. This intuitive sense that the destruction of extraterrestrial life to serve human needs is unethical can be easily dismissed as mere sentimentality, but the fact that so many people hold this view suggests that we need to find the underlying reason why people think this. I would argue that it is not merely a transient fashion or a sentimental feeling that proves to be flawed when subjected to moral scrutiny, but that it reflects a view that living things, with evolutionary trajectories, even if only deterministic and ultimately controlled by genetic information (or its alien equivalent), should be left alone to follow those trajectories in the same way that life on Earth, until the rise of humans, did so. One might even say that this view is in itself embedded in the deeper recognition that we as an intelligent species began our humble beginnings as microbes, and perhaps it is this humility at our own microbial origins that elicits in us a sense of reverence and humility towards other biospheres, even if they are only microbial. It would be easy to dismiss this as trite nonsense, but if nothing else, even if at an absolute minimal level, this attitude stops us from engaging in a wanton extraterrestrial rampage of sampling and destruction of life, motivated only by instrumental needs, an attitude which might have compromised our own biosphere if it had been implemented by some other entity. This is a similar view adopted by Randolph and McKay (2014, 33), who asserted that "intrinsic value provides an important check against humans intentionally or unintentionally skewing their analysis so that promoting human interests is always the most ethical option."

12.5 Planetary Protection

Adjudicating competing interests becomes particularly important when considering planetary protection. These policies concern the contamination of extraterrestrial environments with terrestrial life as well as the contamination of the Earth's biosphere with extraterrestrial life (if it exists) (McKay and Davis 1989; Rummel 2001). In principle, an ethic that encompasses an intrinsic value for microbial life should allow for extraterrestrial exploration. In the same way as the human exploration of Antarctica has not caused widespread destruction of microbial ecosystems there, it is likely that the chance local introduction of organisms on to the surface of another planet would not cause widespread destruction. However, 'likely' does not imply certainty. The chance of the introduced organisms interacting directly with alien hosts, in the same way as a virus does with its host, is highly unlikely as Marshall (1993) recognised. Interactions between hosts and viruses or other parasites are usually highly co-evolved. A more likely route to deleterious contamination would be an introduced organism competing for existing resources or nutrients in an environment, effectively outcompeting other organisms and eventually causing widespread disruption to an alien biosphere. There are reasons to find this unlikely from a microbiological point of view. It would be surprising, for example, if a microbe released from a robotic or human lander was to find extraterrestrial conditions suitable enough to allow for successful reproduction to the extent of outcompeting indigenous life. However, one might imagine hypothetical scenarios, such as the release of a terrestrial organism adapted to polar conditions into Mars-like conditions where it can thrive. There are scenarios one can imagine in which a terrestrial organism could be highly successful in an extraterrestrial environment.

Whatever ethical position we adopt for extraterrestrial life, planetary protection policies provide a useful cautionary backdrop. Harmful contamination that destroys extraterrestrial life is worth preventing in order to avoid the loss of microbes that have instrumental value to humans. If we do assert that they have intrinsic value, then we are motivated to prevent the destruction of extraterrestrial life not merely because we lose instrumental value such as scientific information, but because we seek to prevent the destruction of extraterrestrial life for its own sake, to allow it to fulfil its natural trajectory.

To understand whether the introduction of microorganisms into an alien biosphere would compromise it requires considerable scientific knowledge which includes, but is not limited to, information on the microbial content of spacecraft, knowledge of the potential growth of terrestrial microorganisms in extraterrestrial environments, and knowledge of the extraterrestrial environmental conditions into which life is introduced. Only with this knowledge is it possible to determine which interests should be prioritised: human exploration goals or the local biology.

12.6 Originism

Much of the foregoing makes little consideration of the origin of extraterrestrial life. A fundamental question in any ethics we apply to extraterrestrial life is whether an independent origin of life would command any additional, or less, moral consideration than life related to life on Earth. Does merely being part of another tree of life give life special ethical status?

The question has been met with a variety of different responses, many of them reflecting a view that life of an entirely different origin would command special attention. Lupisella (1997, 89) stated that: "If, however, that life occupies its own separate, unique phylogenetic tree, we may find ourselves asking fundamental questions regarding its status in our world views." Hargrove (1989, 130) asserted that: "I suspect that the organisms would be considered more valuable because they were not part of our system or our history than if they were." Randolph et al. (1997, 1) similarly said that: "From an ethical point of view, the need to preserve a life form, however lowly, must be more compelling if that life form represents a unique life form with an evolutionary history and origin distinct from all other manifestations of life."

Callicott (1989, 262), a strong proponent of Leopold's land ethic, which seeks to site the treatment of the natural world within Leopold's view of the biotic integrity of ecosystems, stated that: "Extraterrestrial life forms, assuming that they were not of Earthly origin and inoculated somehow on some foreign body or vice versa, would not be our kin…Hence they would lie outside the scope of Leopold's land ethic."

In analogy with the word 'speciesism', which refers to the different treatment of life according to which species it belongs to, we might say that this debate revolves around the question of 'originism' Cockell (2010): does being of an independent biochemical origin imply an organism should be treated differently?

We might expect that however alien another life form is, it would still do the things that we associate with biology—reproduce, grow, be subject to evolution, etc.—and therefore it would exhibit such qualities as being 'alive' (i.e. it would fit within existing definitions of life), being sentient or being able to reason. As different schools of ethics are concerned not with specific species but with the characteristics they exhibit, i.e. whether they are alive, sentient or able to reason, then it might be the case that we could formulate an ethic for extraterrestrial life that is consistent with terrestrial ethics and does not take into account the biochemical origin of extraterrestrial life. The chemical structure of the genetic code, or its equivalent, in extraterrestrial life does not seem pertinent to the question of whether it is morally considerable. Therefore, we could reasonably conclude that there is no convincing reason to be originist.

Callicott's view that extraterrestrial life would lie outside of Leopold's Land Ethic because it is not our kith and kin and therefore not part of the terrestrial biosphere is correct, but it would seem egregious to construct an ethic for extraterrestrial life that builds on the foundation of a common evolutionary descent and therefore excludes extraterrestrial life. Fortunately, Leopold never defined the general ethos of the Land Ethic as only working on Earth. Taken in a broader context, we could assert that extraterrestrial life, being presumably embedded with communities of ecosystems of extraterrestrial life, could place demands on us to respect the biotic integrity of those communities as an extraterrestrial extrapolation of the Land Ethic, if indeed the Land Ethic, or a version of it, was a position we chose to adopt.

The term speciesism was made popular by Singer (1975). He was widely misinterpreted by the animal liberation movement as having asserted equal rights for all animals, but his point was that animals deserve equal consideration, not that we would ultimately conclude that all organisms are equally morally significant. The parallels with extraterrestrial life seem clear. We should afford extraterrestrial life equal consideration, simply because we may begin from a position of not understanding its behavioural characteristics and so we will need to understand it before understanding how it fits within ethical systems. From this point of view, its biochemical origin is irrelevant; there is no argument for originism.

However, like different animals, there may well be gradations of sentiency and other characteristics that spring from different biologies and physiologies, characteristics which ultimately affect where in the scheme of life, from microbes to man, different extraterrestrial life forms fit. So like speciesism, there is a good argument to have an originist perspective: different extraterrestrial life forms probably demand different moral significances just as we treat microbes differently to humans.

In summary, there is no ethical position to be originist just because extraterrestrial life is on a different tree of life. Biochemical architecture in itself has no bearing on the ethics we apply to life. However, an originist view of extraterrestrial life does emerge from the observation that extraterrestrial life would likely exhibit behaviours, including degrees and types of sentiency and ability to reason, that we would not fully comprehend without full knowledge of that life. Therefore, from a position of ignorance, we should be originist in giving special treatment to extraterrestrial life in our efforts to understand it and how it might fit within our ethical frameworks.

12.7 Conclusion

The most likely type of life that we will find on other planetary bodies, if we find any at all, is microscopic life. Therefore, our treatment of microbial life on Earth and the ethics we apply to it is likely to be the strongest foundation for understanding how we should treat extraterrestrial life. Here, I have argued that viewing microbial life elsewhere purely from instrumental considerations may be highly destructive and that a form of intrinsic value, rooted in a respect for the trajectories and potential of other life forms and biospheres, allows us to take a more cautious approach to the exploration of other worlds. This ethic is not easy to constrain in a way that allows for clear adjudication in all matters of extraterrestrial exploration, but it prevents a policy of exploration that sees microscopic extraterrestrial life as purely a resource.

References

Attfield, R. (1981). The good of trees. *Journal of Value Inquiry, 15*, 35–54.
Benner, S. A. (2010). Defining life. *Astrobiology, 10*, 1021–1030.

Bentham, J. [1823] (1996). *An introduction to the principles of morals and legislation*. Oxford: Oxford University Press.
Callicott, B. (1986). On the intrinsic value of non-human species. In B. Norton (Ed.), *The preservation of species* (pp. 138–172). Princeton: Princeton University Press.
Callicott, J. B. (1989). Moral considerability and extraterrestrial life. In J. B. Callicott (Ed.), *Defense of the land ethic: Essays in environmental philosophy* (pp. 263–266). Albany: State University of New York Press.
Catling, D. C., Glein, C. R., Zahnle, K. J., & McKay, C. P. (2005). Why O_2 is required by complex life on habitable planets and the concept of planetary 'oxygenation time'. *Astrobiology, 5*, 415–438.
Cleland, C., & Chyba, C. F. (2002). Defining life. *Origins of Life and Evolution of the Biosphere, 32*, 387–393.
Cockell, C. S. (2004). The rights of microbes. *Interdisciplinary Science Reviews, 29*, 141–150.
Cockell, C. S. (2005a). The value of microorganisms. *Environmental Ethics, 27*, 375–390.
Cockell, C. S. (2005b). Planetary protection—A microbial ethics approach. *Space Policy, 21*, 287–292.
Cockell, C. S. (2005c). Duties to extraterrestrial microscopic organisms. *Journal of the British Interplanetary Society, 58*, 367–373.
Cockell, C. S. (2008). Environmental ethics and size. *Ethics and the Environment, 13*, 23–39.
Cockell, C. S. (2010). Ethics and extraterrestrial life. In U. Landfester, N.-L. Remuss, K.-U. Schrogl & J.-C. Worms (Eds.), *Humans in outer space—Interdisciplinary perspectives* (pp. 80–101). Heidelberg: Springer.
Falkowski, P. G. (2015). *Life's engines*. Princeton: Princeton University Press.
Goodpaster, K. (1978). On being morally considerable. *Journal of Philosophy, 22*, 308–325.
Hargrove, E. (1989). *Foundations of environmental ethics*. New Jersey: Prentice-Hall.
Kant, I. (2004). *The metaphysics of morals*. Cambridge: Cambridge University Press.
Landenmark, H. K. E., Forgan, D. H., & Cockell, C. S. (2015). An estimate of the total DNA in the biosphere. *PLoS Biology, 13*, e1002168.
Lupisella, M. (1997). The rights of Martians. *Space Policy, 13*, 89–94.
Lupisella, M. (1999). Ensuring the integrity of possible Martian life. *IAA Paper*, IAA-99-IAA.13.1.08.
Marshall, A. (1993). Ethics and the extraterrestrial environment. *Journal of Applied Philosophy, 10*, 227–236.
McKay, C. P. (1990). Does Mars have rights? An approach to the environmental ethics of planetary engineering. In D. MacNiven (Ed.), *Moral expertise* (pp. 184–197). New York: Routledge.
McKay, C. P., & Davis, W. (1989). Planetary protection issues in advance of human exploration of Mars. *Advances in Space Research, 9*, 197–202.
Persson, E. (2012). The moral status of extraterrestrial life. *Astrobiology, 12*, 976–984.
Race, M. S., & Randolph, R. O. (2002). The need for operating guidelines and a decision making framework applicable to the discovery of non-intelligent extraterrestrial life. *Advances of Space Research, 30*, 1583–1591.
Randolph, R. O., & McKay, C. P. (2014). Protecting and expanding the richness and diversity of life, an ethic for astrobiology research and space exploration. *International Journal of Astrobiology, 13*, 28–34.
Randolph, R. O., Race, M. S., & McKay, C. P. (1997). Reconsidering the theological and ethical implications of extraterrestrial life. *CTNS (Center for Theology and the Natural Sciences, Berkeley) Bulletin, 17*, 1–8.
Regan, T. (1983). *The case for animal rights*. Berkeley: University of California Press.
Rolston, H. (1988). *Environmental ethics: Duties to and values in the natural world*. Philadelphia: Temple University Press.
Ruiz-Mirazo, K., Paretó, J., & Moreno, A. (2004). A universal definition of life: Autonomy and open-ended evolution. *Origins of Life and Evolution of the Biosphere, 34*, 323–346.

Rummel, J. D. (2001). Planetary exploration in the time of astrobiology: Protecting against biological contamination. *Proceedings of the National Academy of Sciences, 98*, 2128–2131.

Sagan, C. (1980). *Cosmos*. New York: Random House.

Singer, P. (1975). *Animal liberation*. New York: Avon Books.

Smith, K. (2009). The trouble with intrinsic value: An ethical primer for astrobiology. In C. M. Bertka (Ed.), *Exploring the origin, extent, and future of life: Philosophical, ethical, and theological perspectives* (pp. 261–80). Cambridge: Cambridge University Press.

Sturba, J. (1995). From biocentric individualism to biocentric pluralism. *Environmental Ethics, 17*, 191–207.

Taylor, P. (1989). *Respect for nature: A theory of environmental ethics*. Princeton: Princeton University Press.

Varner, G. (2002). Biocentric individualism. In D. Schmitz & E. Willott (Eds.), *Environmental ethics: What really matters, what really works* (pp. 108–120). Oxford: Oxford University Press.

Whitman, W. B., Coleman, D. C., & Wiebe, W. J. (1998). Prokaryotes: The unseen majority. *Proceedings of the National Academy of Sciences, 95*, 6578–6583.

Chapter 13
Kantian Foundations for a Cosmocentric Ethic

Anna Frammartino Wilks

13.1 Getting a Cosmocentric Ethic off the Terrestrial Ground

The deeply perplexing issues arising from current planetary exploration are signaling that we need more than just ingeniously designed spacecraft before launching into extraterrestrial territory. We need in addition to equip ourselves with a *cosmocentric ethic* adequately tailored to the distinct ethical entanglement that emerges from our efforts to enlarge the circumference of our own world by transforming others. Professionals and researchers from a variety of disciplines and agencies are voicing this need.[1] Robert Haynes asserts that: "We need from philosophers a new 'cosmocentric' ethics, and perhaps a revised theory of intrinsic value" to achieve such breadth (1990, 177). Mark Lupisella echoes this sentiment with a particular emphasis on the latter point in his assertion that "an important challenge for a cosmocentric ethic is justifying intrinsic value" (2009a, 193) . I attempt here to take on this challenge by offering an account of *intrinsic value* sufficiently broad enough to address ethical issues on a cosmic scale.

My account is rooted in some central precepts of the philosophy of biology and ethical theory of Immanuel Kant.[2] The cosmocentric ethic I develop, however, is

[1]McKay (1990, 186), Lupisella and Logsdon (1997).
[2]Citations from the works of Immanuel Kant include the following abbreviated English titles: *CPR* (*Critique of Pure Reason*), *CPJ* (*Critique of the Power of Judgment*), *OP* (*Opus Postumum*), *GMM* (*Groundwork of the Metaphysics of Morals*), *First Introduction* (*First Introduction to The Critique of the Power of Judgment*). They also include the corresponding volume and page numbers of the standard *Akademie* edition: *Kants gesammelte Schriften*. Citations from the *Critique of Pure Reason*, however, are indicated by the standard pagination of the A and B edition. All quotations are taken from the English translation of Kant's texts in Guyer, Paul and Wood, Allen W. (eds), *The Cambridge Edition of the Works of Immanuel Kant*. Cambridge: Cambridge University Press, 1992.

A.F. Wilks (✉)
Acadia University, Wolfville, Canada
e-mail: anna.wilks@acadiau.ca

J.S.J. Schwartz and T. Milligan (eds.), *The Ethics of Space Exploration*, Space and Society, DOI 10.1007/978-3-319-39827-3_13

not intended to be representative of Kant's own ethical position. Nonetheless I think Kant's account of biological beings highlights a rudimentary feature which, when sufficiently developed, gives rise to moral agency. Living beings, Kant maintains, are regarded *as if* they are capable of a kind of *self-organization* that is irreducible to mechanical explanation. Moral agents, on Kant's view, are characterized by their capacity for *self-legislation*—the feature that endows them with *intrinsic value*. My aim is to show that the capacity for self-legislation possessed by moral beings is just a heightened or more pronounced instance of the general capacity for self-organization manifested by *all* living beings. I conclude from this that intrinsic value is not, therefore, something that may only be attributed to self-legislating beings but rather something that can be extended to *all* living beings. I then proceed to show that, nonetheless, there is still a non-arbitrary basis for viewing self-legislating beings as possessing *greater* intrinsic value than non-self-legislating beings. Equipped with a more clearly defined account of intrinsic value, we may stand a better chance of establishing the moral bounds of the universe.

While the issues pertaining to the ethics of space exploration are vast, I limit myself here to the problem of determining appropriate ethical principles to govern our interaction with primitive indigenous life forms that we may encounter in our efforts to engineer extraterrestrial ecosystems or terraform other planets (Schwartz 2015). Current plans for exploration on Mars present us with the prospect of possibly harming or destroying any extant life forms there, or affecting the potential for future Martian life by adversely altering the Martian biosphere. Can such primitive life forms be considered to possess any value? If so, do they possess merely instrumental or intrinsic value? Biocentrists such as Charles Cockell defend a view that attributes rights to microbes, and one such right is to be left to develop and evolve undisturbed by other life forms, including humans (Cockell 2004, 2005). But what is the basis for these rights? Unless rooted in an adequate theory of *intrinsic value*, such claims are unfounded.[3]

In my attempt to provide an account of intrinsic value that meets the demands of a cosmocentric ethic, I limit myself to the treatment of arguments based on the intrinsic rather than instrumental value of extraterrestrial life—though the latter are certainly important for dealing with other space exploration issues.[4] Ultimately, I contend that: The view that it is morally impermissible to terraform other planets inhabited by indigenous life forms cannot be defended solely on the biocentric principle that "all living things have intrinsic value."[5] Moreover, I support this position with a cosmocentric ethic that is more closely aligned with biocentrism than with anthropocentrism. My strategy is to avail myself of some central precepts in the ethical and biological theory of Immanuel Kant. My approach, however,

[3]More recently, Cockell (2011) has appealed to Aristotelian teleology to defend the intrinsic value of microbial life. See also Cockell's contribution to this volume.

[4]I share the majority view here that extraterrestrial microbes possess significant instrumental value.

[5]I acknowledge, however, that there may be other valid reasons for refraining from such practices. Typically, however, these other reasons are fueled by anthropocentric interests.

departs from the one typically adopted by adherents of Kant's position in these debates, as it emphasizes the biocentric rather than ratiocentric strains in Kant's thought, which, however, Kant himself leaves undeveloped. In the section that follows, I lay out the key elements that I think constitute the general parameters for a cosmocentric ethic, and in the section after that I show how my account of intrinsic value fits into that ethic.

13.2 Determining the Cosmocentric Element in a Cosmocentric Ethic

The very idea of a *cosmocentric ethic* seems to suggest an ethic that prioritizes the cosmos. But how exactly is such an idea to be construed? Our lack of fundamental knowledge concerning the extent and ultimate nature of the universe presents serious challenges. Moreover, many argue that moral laws stem from the viewpoint of human beings—beings capable of forming moral values—and that therefore, moral values and principles have application only for and between human beings. Some appeal to this precept to support *anthropocentrism*—an ethical framework that prioritizes *human beings*.[6] Others, like Don MacNiven, suggest rather that we need to *transcend* our human perspective in order to acknowledge the interests and values of non-human beings (MacNiven 1990). Non-anthropocentrists such as Paul Taylor, Holmes Rolston, and J. Baird Callicott argue that such transcendence is impracticable, though it is nonetheless possible to *extend* moral consideration to non-human beings and ecosystems (Hargrove 1992, 184).

While a *cosmocentric ethic* may very well have to be based on a human viewpoint informed by human values, it may nonetheless be possible to establish a criterion of intrinsic value for universal application beyond the terrestrial realm. The notion of *intrinsic value* that I adopt is this: *a value that something possesses by virtue of its intrinsic properties or nature.* I maintain that establishing a criterion of intrinsic value for a cosmocentric ethic primarily requires the specification of a fundamental property that may, in principle, be extended to any being in the universe. Such a property could then serve as a non-arbitrary criterion for determining which beings possess intrinsic value by assessing the degree to which they manifest this fundamental property (Lupisella and Logsdon 1997, 1).[7] Thus a cosmocentric ethic would require us to consider the realm of candidates to be *all* the beings in the universe—even in the absence of knowledge concerning what *particular* things the universe contains. This is possible on my account because the criterion I adopt is a purely formal one and for this reason does not require a catalogue of the totality of

---

[6]Schwartz (2013, 9–11) provides an insightful account of some central anthropocentric arguments concerning these issues.

[7]This is the strategy suggested by Lupisella and Logsdon for the establishment of a cosmocentric ethic.

existing beings in the universe, or a complete account of the nature of the universe. Because of the formal nature of my criterion, the position I defend may be viewed as a *weak* cosmocentric ethic. It is *cosmocentric* in the sense that it gives priority to *all beings in the universe that possess intrinsic value*, appreciating that intrinsic value is not restricted to terrestrial beings; it is *weak* in the sense that it does not attribute intrinsic value to *the universe itself* as a being detached from, or transcending, the beings that constitute the universe.[8] For this reason, my position does not amount to a "boot-strapped" view according to which the universe itself actually acquires intrinsic value *through* certain beings contained in it, which themselves possess intrinsic value (Lupisella 2009b, 333). With these constraints in place, I now proceed to my account of a cosmocentric ethic.[9]

While acknowledging the epistemological constraints that Kant specifies concerning our understanding of the metaphysical nature of the universe, it is, I think, legitimate to establish some claims about our *experience* of the beings in the universe, which may help achieve the *practical* aim of providing a guiding principle for a cosmocentric ethical theory.[10] Most notably, the universe manifests the production of a complexity of diverse beings that are organized into structured unities as formed entities or integrated systems. Among these various organized entities are beings that possess the source of this organization or integration within themselves, i.e., they are *self-organizing*. In this sense, such beings may be viewed as *autonomous*. The property of *self-organization*, I maintain, is the fundamental property that grounds our intuitions about intrinsic value. That is to say, of all the properties possessed by the universe's known inhabitants, self-organization is the property that constitutes the basis for intrinsic value. This claim is to be taken as axiomatic in my argument, though in the sections that follow, I show why this is warranted.

The idea of self-organization has some affinity with the idea of *complexity* that certain thinkers attribute to the development of the universe through evolution and the laws of physics.[11] Lupisella remarks, however, that the criterion of complexity alone might not be sufficient to constitute a basis for the notion of *intrinsic value* that could properly inform a cosmocentric ethic. He recommends that the criterion of complexity be supplemented by other factors. I endorse this recommendation, and suggest that it is not mere complexity that constitutes the basis for intrinsic value, but an *autonomously ordered complexity*, which is what is involved in *self-organization*. Martyn J. Fogg intimates that the universe *itself* manifests the capacity for autonomous action captured in my notion of *self-organization* (2000, 210). Although my own argument does not require that "the universe itself" possess

---

[8]I am grateful to James Schwartz for his suggestions in working out a formulation of this point.

[9]I endorse here Kant's rejection of the possibility of a "rational cosmology" (*CPR* A508/B536-A532/B560).

[10]See Guyer (2005), Ginsborg (2006).

[11]Davies (1995, 102), Smith (2009).

this property, it also does not preclude this possibility—provided that we could form an adequate concept of "the universe itself" in the first place.

Thus I maintain that for the practical purpose of establishing the grounds of a cosmocentric ethic, the basis of *intrinsic value* is the fundamental property of *self-organization*. I argue from this initial axiom that: *the extent to which any being in the universe possesses intrinsic value is determined by its capacity for self-organization*. The issue that confronts us now is: What kinds of beings manifest the capacity for self-organization? I probe this issue in the section that follows.

13.3 The Intrinsic Value of All Living Beings

When considering the various kinds of existing beings that we know of, there seems to be a powerful tendency to view *living beings* as constituting a distinct category. There is just something about living beings that strikes us as unique—though what exactly it is quite eludes us (Cleland 2012). It was the acknowledgment of this fact that inspired the initiative among late eighteenth century scientists to establish the sub-discipline of "biology" as a scientific inquiry specifically devoted to the investigation of living things (Van den Berg 2014). Kant's characterization of a *living thing* was a catalyst in moving this initiative forward, serving as the foundational principle for the life sciences in this early phase, and it is still operative in current biological theory (Deacon and Cashman 2013). What was the distinguishing feature of living things that Kant fastened on?

According to Kant, the fundamental feature of living things that take the form of organisms is that they are *self-organizing* (*CPJ* 5: 372). An organism is not only an *organized* being, as in the case of a machine; it is a *self-organizing* being. We conceive of an organism *as if* it were not controlled by something external to itself, but rather something internal. Thus an organism appears, in this sense, to be a *system* rather than an aggregate, and a somewhat autonomous agent. As such, an organism is conceived *as if* it were a *natural purpose,* since it manifests *reciprocal causality*, i.e., its parts exist for the sake of the whole, and the whole exists for the sake of its parts (*CPJ* 5: 371).[12] The organism does not exist solely to serve some extrinsic purpose (*CPJ* 5: 374). It seems, rather, to contain an *intrinsic purpose* in accordance with which it directs all its activities. All other fundamental features of organisms, e.g., self-maintenance, growth, homeostasis, metabolism, reproduction, evolution through adaptation, etc., depend upon the organism's capacity for self-organization.

The only way we can make sense of the organism's activity is to consider it *as if* it contained a purposive drive within itself. Thus Kant characterizes an organism as

[12]Kant's position should not be confused with that of Aristotle who defends a genuine teleology involving real ends and purposes in nature. The "*as if*" qualification is indicative of Kant's more heuristic sense.

a *natural purpose* (*Naturzweck*).[13] It is this feature of organisms, Kant maintains, that renders a completely mechanical explanation of them impossible, since an organism is always underdetermined by the laws of physics.[14] In effect "self-organization implies intrinsic purposiveness" (Thompson 2010, 140). My view is that the purposiveness exemplified by living organisms is a rudimentary form of a moral constraint. Hannah Ginsborg intimates this view in her claim that Kant's notion of *end* or *purpose* here is bound up with the notion of *normativity*. She maintains that to view an organism as having an end or purpose is essentially "to regard it as conforming to, and a fortiori as governed by, normative rules or constraints" (2006, 464). Kant explains that in our judgment of organisms we compare "the concept of a product of nature as it is with one of what it ought to be" (*First Introduction* 20: 240). I maintain that this "natural normativity," as Ginsborg calls it, may be construed as a prototype of moral constraint that generates a primitive form of moral agency. This is the feature that distinguishes organisms as unique types of beings,[15] and which in some highly developed organisms gives rise to fully autonomous moral agents.

To be sure, since Kant's time the notion of "self-organization" has acquired a much broader meaning, which opened the door to its application to numerous other contexts and fields besides living organisms and biology (Keller 2008). Those applications of the term, however, fail to capture the particular feature of "self-organization" that warrants its association with the domain of ethics, and specifically, for determining the basis of intrinsic value. Kant's characterization of organisms provides us with a *paradigm case* of self-organization for assessing intrinsic value (*CJP* 5: 374–6).[16] Merely existing beings do not possess this property. Thus, on my view, MacNiven's attribution of intrinsic value to systems of merely *existing* beings—which I call an *ontocentric ethic*—would be unfounded. The abiotic features of ecosystems also do not possess this property.[17] It is only by virtue of the organic beings in an ecosystem that the ecosystem itself may be said to involve self-organization. I conclude, therefore, that all living beings, understood as organisms in the Kantian sense, may be viewed as possessing intrinsic value by virtue of their capacity for self-organization.

---

[13]Although Kant's notion of a "natural purpose" is merely to be taken as a *regulative* concept, as opposed to a *constitutive* concept (Kant *CPJ* 5: 375), it nonetheless points the way to establishing the sort of grounds required to warrant viewing something as an end, and not merely as a means, i.e., as something possessing intrinsic rather than merely extrinsic value. See also Guyer (2005).

[14]More accurately, organisms manifest "a kind of mechanism that is subordinated to teleology" (Ginsborg 2006, 462).

[15]Though Kant's anti-reductionist account of organisms is still unpopular with many, anti-reductionism seems to be undergoing somewhat of a revival in current biology. See Henning and Scarfe (2013).

[16]Kant himself, however, did not make this move.

[17]Schwartz (2013, 22) raises additional compelling arguments against preservationists such as Holmes Rolston, Keekok Lee, Alan Marshall and Robert Sparrow, and expresses objections to the arbitrary "postulation of intrinsic values." I attempt to offer here a method for the non-arbitrary selection of such values.

However, the fact that all living beings possess intrinsic value does not necessarily imply that they all possess *equal* intrinsic value, as Taylor's biocentric ethic presupposes.[18] Some maintain that intrinsic value is an all or nothing thing; it does not admit of degrees.[19] I contend that intrinsic value *does* admit of degrees, and that assessing the degree of intrinsic value possessed by living beings is precisely what is required to resolve some of the most important questions pertaining to the ethics of space exploration, in particular, the question of how the value of microbial extraterrestrial life is to be weighed against the value of human life, along with human desires and activities.[20] I direct my attention to the issue of *assessing intrinsic value* in the section that follows.

13.4 The Distinction Between Self-organizing and Self-legislating Beings

I contend that within the category of *intrinsic value* that can be attributed to things there is a continuous *spectrum*. In other words, intrinsic value admits of *degrees*. I have argued that the relevant trait possessed by living beings is self-organization, a manifestation of their apparent autonomy. Thus the degree to which something possesses intrinsic value is determined by the degree to which it is self-organizing. The task that now confronts us is how to assess the degree to which something is self-organizing.

It is in tackling this task that we must transition from biology to ethics. At this juncture as well, a Kantian framework offers guidance. On Kant's view, certain living beings are distinguished from other living beings by virtue of their capacity to legislate to themselves the moral law of reason, which enables them to function as *moral agents*. This capacity renders rational beings *self-legislating*. Because the moral law of reason is one in form, abiding by this law permits a being to achieve *unity* of person and character by virtue of their more robust autonomy, thus fortifying the being's self-organization.[21] Consequently, beings that are capable of self-legislation are more self-organizing than beings that are not—since self-legislating beings are more autonomous and greater autonomy implies greater self-organization.[22] It is legitimate, therefore, to view self-legislating beings as possessing greater intrinsic value than beings that are merely self-organizing.

[18]Taylor (2011) defends "species egalitarianism."

[19]Smith (2009, 269) defends the view that there are distinct categories for determining whether or not a being has intrinsic value. I suggest that there is a continuum, without clearly distinct boundaries.

[20]Schmidtz (1998) and Milligan (2015, 120) also challenge the species egalitarianism claim.

[21]Korsgaard (1999) refers to this self-organization as "self-constitution."

[22]Kant (*GMM* 4: 440–441). For Kant *autonomy* does not mean "doing what one wants" but rather "doing what the moral law" demands. The former is *heteronomy*, which has nothing to do with morality.

To avoid a serious misinterpretation of my position, it is paramount to acknowledge that I consider all beings of a given species to possess *equal* intrinsic value.[23] Moreover, when assessing their capacity for self-organization, all that matters is the potential for some members of the species to manifest that capacity in the current phase of their evolutionary history. It is not necessary for every member to manifest that property. This crucial aspect of my theory precludes the possibility of employing my argument to discriminate among the members within a species on the basis of race, gender, education, cultural background, wealth, religion, political orientation, sexual orientation, etc. My position, in fact, entails the *prohibition* of exploitation or oppression of some individuals over others in a given species. The above constraint also ensures the extension of the degree of intrinsic value to young, not fully developed, physically impaired, or cognitively impaired members of a given species. However, although the property of intrinsic value may be extended to all the members of a species just by virtue of *some* members possessing a certain degree of self-organization, this does not entail that *every* property of some individuals pertains to all the individuals in the species, e.g., moral responsibility. My claim is simply that all members of the species are equal in terms of their status as *moral patients*, rather than *moral agents*.

This theory also precludes the arbitrary supremacy of the anthropocentric viewpoint, as the only crucial factor determining intrinsic value is the capacity for self-organization, which is not a capacity limited to human beings, or even rational beings. Displacing the anthropocentric viewpoint from the centre of my cosmocentric ethic has the advantage of extending intrinsic value to all living beings in the universe; at the same time, the acknowledgment that self-organization admits of degrees enables us to attribute, in a non-arbitrary manner, the appropriate *degree* of intrinsic value pertaining to a given species.[24] In addition, a significant difference between Kant's view and my own is that Kant asserts that being self-legislating is both a necessary and sufficient condition for the possession of intrinsic value. On my view, being self-legislating is merely a sufficient but not a necessary condition for intrinsic value. I maintain that *capacity for self-organization* is the only necessary and sufficient condition for intrinsic value. Consequently, intrinsic value may be attributed to all living beings.

I contend, therefore, that in a genuinely cosmocentric ethic, the system of *all living beings in the universe* constitutes the domain of beings possessing intrinsic value, by virtue of their being self-organizing beings. Furthermore, there is a basis for attributing greater intrinsic value to some species of living beings over others, depending on the *degree* of self-organization they possess. As argued above, the possession of the capacity for *self-legislation* is a heightened or more pronounced form of self-organization, as it is indicative of a more genuine form of autonomy.

---

[23] See Gilbert et al. (2012) for discussion on the debate over the ontological status and demarcation of biological individuals and species.

[24] Smith (2014, 210) adopts a different strategy to displace the anthropocentric viewpoint from the centre of a cosmocentric ethic by appealing to a *ratiocentric* principle, supported by his concept of the "sociality-reason-culture" triad.

Thus a species that possesses the capacity for self-legislation possesses greater intrinsic value than a species that lacks this capacity. In general, the more a species achieves the capacity for self-legislation, the greater its intrinsic value.

13.5 The *Respect for Intrinsic Value* Test

To facilitate the implementation of this cosmocentric ethic I propose two things: (a) a fundamental *principle* for determining the degree of *intrinsic value* of living beings, and (b) a *test* for the practical application of this principle. We begin with the biocentric principle established in the previous sections: All living beings possess intrinsic value. Furthermore, all members of a given species possess *equal* intrinsic value. For example, all human beings have equal intrinsic value; all hedgehogs have equal intrinsic value; all beta fish have equal intrinsic value, etc.

The principle I offer is this: The degree of intrinsic value that an intrinsically valuable being possesses is proportional to the degree of its capacity to respect the intrinsic value of other intrinsically valuable beings. Consequently, an intrinsically valuable being that has the capacity to respect the intrinsic value of other intrinsically valuable beings possesses greater intrinsic value than a being that lacks this capacity.[25]

The test I offer for the application of this principle is what I call the *Respect for Intrinsic Value Test* (RIV test). This test is grounded in the Kantian precept that a being with intrinsic value is one that is capable of "respect for the moral law" by means of its freely *legislating* this law to itself—the factor that constitutes its autonomy. Specifically, the moral law dictates that beings with intrinsic value must always be treated as ends in themselves and never as a means only. The test is, in substance, my articulation of a modified version of Kant's *categorical imperative* (*GMM* 4: 420–441).

The test works like this: If an intrinsically valuable being x has the capacity to respect the intrinsic value (IV) of another intrinsically valuable being y, but y does not have the capacity to respect the IV of x, then x thereby manifests greater IV than y. Consequently, if for example, human beings have the capacity to respect the IV of certain Martian microbes, but the Martian microbes do not have the capacity to respect the IV of the human beings, the human beings thereby manifest greater IV than the microbes, by virtue of the microbes' having performed less well than the human beings on the RIV test. On the other hand, if *both* the human beings and the microbes perform *equally* well on the RIV test, then the original equality of their capacity for self-organization is preserved, and thus their original intrinsic value is preserved. To be sure, evaluating a being's performance on the RIV Test poses

[25]Smith (2014, 211) suggests that the "sociality-rationality-culture triad" constitutes the basis of intrinsic value. My position may be construed as a development of Smith's view in that the principle I offer isolates the *catalyst* that initiates the dynamics of this triad.

significant epistemological challenges in many cases. For example, what would count as a *manifestation* of "respect for intrinsic value" in non-human beings, and by what means are we to assess their intentionality and their conscious awareness of value?[26] While these deeply complex epistemological issues merit considerable attention, an adequate treatment of them would take us far beyond the scope of this paper. The central point, however, is that the RIV test is not a "pass or fail test" but rather a "performance test," i.e., a test designed to rate a thing's performance relative to other things, rather than to include it or exclude it from some category. In this sense the test is intended to remove an imagined boundary between things possessing intrinsic value and things which completely lack intrinsic value, by requiring us to acknowledge that intrinsic value may be possessed in degrees.

Equipped with a formal principle of intrinsic value, and with a practical test for its application, we may now consider how various species of living beings would most likely fare in their performance on the RIV test. Clearly, a species of completely self-legislating beings would do extremely well, since they would possess the greatest capacity for respecting the IV of other self-organizing beings. In fact such beings may, for this reason, be viewed as possessing greater intrinsic value than *all* other species of living beings. It is clear that human beings are capable of at least *some* degree of self-legislation, and would thus undoubtedly score quite high. I do not, however, think that it is necessarily the case that *only* human beings would be a species of this sort; the universe may contain other kinds of non-human self-legislating beings.

Moreover, because different species of living things may achieve the capacity for self-legislation in varying degrees, it is also reasonable to suppose that other kinds of living beings that *approximate* the capacity for self-legislation would perform *somewhat* well. For example, species that are sentient, e.g., dogs, dolphins, birds, etc., might demonstrate the capacity to respect the intrinsic value of at least *some* other members of their own species through their ability to care for their offspring. The next level down might include living beings manifesting some degree of sociality, e.g., ants and bees. Among the kinds of living beings who would score even lower on the RIV test would be those that only manifest the capacity to respect their *own* intrinsic value—manifested by their efforts in self-preservation—but not that of any other. Examples of these kinds of living beings would most probably include microbes and plants. It should be noted, furthermore, that one may not argue that a species that currently lacks the actual capacity to perform well on the test may possess the potential to *evolve* into a species that may perform well on the test at some future point. The reason is that, as stated earlier, at least *some* members of the species must *actually* possess the given capacity in the *current* phase of the species' evolutionary history.

This method of assessing intrinsic value addresses Fogg's concerns that prohibiting human beings from terraforming Mars would amount to the view that "humans actually have the *lowest* degree of intrinsic worth of any class of formed

[26] I thank James Schwartz for raising these important issues.

object" (2000, 210),[27]despite the fact that the culture and technology of human beings also arose from biological evolution and are thus themselves *natural.* The RIV test would ensure that human beings, along with all other living beings may be assured a non-arbitrary assessment of their intrinsic value. Thus, provided there were adequate reasons, my position could conceivably justify activities that might result in harm or destruction to extraterrestrial microbial life forms by other life forms such as human beings, provided again that those microbes do not perform as well as human beings on the RIV test. It should be noted, however, that by the same token, if human beings should encounter extraterrestrial life forms that exceed their own performance on the RIV test, then those beings would thereby possess greater intrinsic value than human beings, and human beings would be in a position of subordination to *them* (Kant, *OP* 21: 215).[28]

My argument, however, requires that, regardless of which species of living beings manifests greater intrinsic value, their treatment of living beings should always acknowledge the appropriate degree of intrinsic value of those beings. In the event that the more valuable species fails in this, and that they instead engage in disrespectful behaviour such as wanton destruction or exploitation, then such behavior would be indicative of the fact that they do not, in actual fact, possess greater intrinsic value than the beings they exploit. No doubt, there would be numerous factors that would enter the picture to complicate the assessment of intrinsic value in accordance with the RIV test I propose. Thus a reasonable assessment would, in the end, require an appeal to "practical wisdom" as suggested by Tony Milligan in his expression of skepticism concerning the adequacy of an algorithm for assessing value (2015). While an algorithm such as the RIV test I suggest will not do the whole job, it does do an important part of it. It provides a basis for the presupposition that human beings may very well have a duty to extend human life. In fact, Milligan himself acknowledges that this presupposition requires defense (2015, 58). The RIV test provides a non-arbitrary means for providing this defense.

Though the RIV test aims at achieving only rough assessments of intrinsic value, it has, I think, greater practical import than Rolston's overly simplistic dichotomy between *holders of value* and *beholders of value* (Hargrove 1992, 193). In fact, the RIV test illustrates how being a *beholder of value* actually contributes to one's status as a *holder of value*. Another notable advantage of the RIV test is that any IV *inequality* that results is something that is objectively *established*, by means of the participants' performance on the test. The *degree* of IV is something that the participants may acquire either more or less of—depending, to a very significant extent, on their own test performance. IV inequality is not arbitrarily assumed.

[27]See also Zubrin and Wagner (1996, 248–249).

[28]In fact, Kant himself conceived of the possibility of living beings more perfect than human beings. See also Baum (2010) and Baum's contribution to the present volume.

Furthermore, it should be stressed that a participant's acquisition of greater IV than another participant on this test does not *negate* the IV of the latter, and transform it into mere instrumental value. Rather, the latter's IV is still preserved, only it is now diminished in *degree* by its having performed less well on the RIV test than the other participant. Moreover, although the RIV test does not enable one to quantify the precise degree of IV in any possible situation that may arise, it does provide some general guidelines as to how we may *begin* to weigh the IV of different kinds of living beings in a non-arbitrary manner.

13.6 Concluding Remarks

My cosmocentric ethic provides a starting point for addressing, in particular, the issue of how to assess, in a non-arbitrary fashion, the relative value of different kinds of living beings, all of which, by virtue of being living beings, possess some degree of intrinsic value. This has significant utility in attacking the problem of the moral permissibility of conducting activity on other planets that could disrupt the indigenous microbial life on those planets. I conclude thus: *The view that "it is morally impermissible to terraform other planets inhabited by indigenous life forms" cannot be defended solely on the biocentric principle that "all living things have intrinsic value."* As I have shown, the moral impermissibility of such activity requires a compelling argument in support of the much stronger view that: "*all living things have* equal *intrinsic value*"—which I deny. Thus, provided space exploration is conducted with a genuine respect for the *intrinsic value* of possible indigenous extraterrestrial life forms, it is not, in principle, morally impermissible to prioritize the value of human beings in such endeavours, given their considerably greater degree of intrinsic value than microbial beings, by virtue of their capacity for *self-legislation*.[29]

Nonetheless, this does not grant free reign to human beings in their treatment of less intrinsically valuable life forms, including microbes, since, as living beings, they still possess some *degree* of intrinsic value. I am merely arguing that our moral obligations to them ought to be determined in consideration of the intrinsic value of *other* living beings—especially those possessing greater intrinsic value. Furthermore, given that, on this view, all life forms have intrinsic value, and that life has value and priority over non-life, I agree with Christopher McKay that it is morally permissible to undertake technological endeavours both (a) to protect and promote the survival, richness and diversity of indigenous, extraterrestrial life forms

[29]In addition, there may be strong reasons based on instrumental values for refraining from numerous kinds of activities pertaining to space exploration. I have only limited myself here to arguments that appeal to reasons based on intrinsic value.

on other planets, and also (b) to create an extraterrestrial biosphere that could generate and sustain life on planets that do not currently have life (1990, 194).

I have attempted to offer here an account of intrinsic value to address the demands of a weak cosmocentric ethic. My hope is that the deficiencies in this account may inspire continuing reflection on this important issue, and propel us further towards the establishment of a fully developed *cosmocentric ethic* that will put the human exploration of space in its place.[30]

References

Baum, S. (2010). Universalist ethics in extraterrestrial encounter. *Acta Astronautica, 66*, 617–623.

Cockell, C. S. (2004). The rights of microbes. *Interdisciplinary Science Reviews, 29*, 141–150.

Cockell, C. S. (2005). The value of microorganisms. *Environmental Ethics, 27*, 375–390.

Cockell, C. S. (2011). Ethics and extraterrestrial life. In U. Landfester, N.-L. Remuss, K.-U. Schrogl, & J.-C. Worms (Eds.), *Humans in outer space—Interdisciplinary perspectives* (pp. 80–101). Vienna: Springer.

Davies, P. (1995). *Are we alone? Philosophical implications of the discovery of extraterrestrial life*. New York: Harper Collins Publishers.

Deacon, T., & Cashman, T. (2013). Teleology versus mechanism in biology: Beyond self-organization. In B. G. Henning & A. C. Scarfe (Eds.), *Beyond mechanism putting life back into biology* (pp. 287–308). Lanham: Lexington Books.

Fogg, M. J. (2000). The ethical dimensions of space settlement. *Space Policy, 16*, 205–211.

Gilbert, S. F., Sapp, J., & Tauber, A. (2012). A symbiotic view of life: We have never been individuals. *The Quarterly Review of Biology, 87*, 325–341.

Ginsborg, H. (2006). Kant's biological teleology and its philosophical significance. In G. Bird (Ed.), *A campanion to kant* (pp. 455–469). Malden: Blackwell Publishing.

Guyer, P. (2005). *Kant's system of nature and freedom*. Oxford: Oxford University Press.

Guyer, P., & Wood, A. W. (Eds.). (1992). *The cambridge edition of the works of Immanuel Kant*. Cambridge: Cambridge University Press.

Hargrove, E. C. (1992). Weak anthropocentric intrinsic value. *The Monist, 75*, 183–207.

Haynes, R. H. (1990). Ecce ecopoiesis: Playing god on mars. In D. MacNiven (Ed.), *Moral expertise* (pp. 161–183). London: Routledge.

Keller, E. F. (2008). Organisms, machines, and thunderstorms: A history of self-organization, part one. *Historical Studies in the Natural Sciences, 38*, 45–75.

Korsgaard, C. M. (1999). Self-constitution in the ethics of Plato and Kant. *The Journal of Ethics, 3*, 1–29.

Lupisella, M. L. (2009a). Cosmocultural evolution. In S. Dick & M. L. Lupisella (Eds.), *Cosmos and cultural evolution in a cosmic context* (pp. 321–359). Washington D.C.: National Aeronautics and Space Administration.

Lupisella, M. L. (2009b). The search for extraterrestrial life: Epistemology, ethics, and worldviews. In C. M. Bertka (Ed.), *Exploring the origin, extent, and future of life* (pp. 186–204). Cambridge: Cambridge University Press.

Lupisella, M. L., & Logsdon, J. (1997). Do we need a cosmocentric ethic? *International Astronautical Federation Congress*, (pp. 1–10). Turin: American Institute of Aeronautics and Astronautics.

MacNiven, D. (Ed.). (1990). *Moral expertise*. London: Routledge.

[30]I thank Ford W. Doolittle and Carlos Mariscal for their helpful discussions.

McKay, C. P. (1990). Does Mars have rights? In D. MacNiven (Ed.), *Moral expertise* (pp. 184–197). London: Routledge.

Milligan, T. (2015). *Nobody owns the moon: The ethics of space exploitation*. North Carolina: McFarland and Company Publishers.

Schmidtz, D. (1998). Are all species equal? *Journal of Applied Philosophy, 15*, 57–67.

Schwartz, J. S. J. (2013). On the moral permissibility of terraforming. *Ethics and the Environment, 18*, 1–31.

Schwartz, J. S. J. (2015). Philosophy and Space Exploration Bibliography [online August version].

Smith, K. C. (2009). The trouble with intrinsic value: An ethical primer for astrobiology. In C. M. Bertka (Ed.), *Exploring the origin, extent, and future of life* (pp. 261–280). Cambridge: Cambridge University Press.

Smith, K. C. (2014). Manifest complexity: A foundational ethic for astrobiology. *Space Policy, 30*, 209–214.

Taylor, P. W. (2011). *Respect for nature: A Theory of Environmental Ethics*. Princeton: Princeton University Press.

Thompson, E. (2010). *Mind in life, biology, phenomenology and the sciences of mind*. Cambridge: Harvard University Press.

van den Berg, H. (2014). *Kant on proper science*. Dordrecht: Springer.

Zubrin, R., & Wagner R. [1996], (2011). *The case for mars: The plan to settle the red planet and why we must*. New York: Free Press.

Chapter 14
The Curious Case of the Martian Microbes: Mariomania, Intrinsic Value and the Prime Directive

Kelly C. Smith

14.1 An Introduction to Mariomania

Astrobiology provides new vistas in which to consider whether and how we should extend moral consideration to non-human entities. Should we discover microbes on Mars, it is unclear how our obligations to them would differ from those we have towards terrestrial microbes. Certainly if such organisms represent an independent origin of life, they would provide a treasure trove of scientific information about the evolutionary origins of life—one of the biggest mysteries confronting modern science. Dissimilarity cuts both ways, though, and the fact that Martian microbes don't share an ecosystem with terrestrial organisms undercuts one major kind of argument used in terrestrial environmental ethics.[1] So what precisely are our obligations to Martian life?

We are now confident that Martian life, if it exists at all, will be strictly microbial in nature (Mckay 1990; Gibson et al. 1997). This further complicates the moral question, since many of the classic debates in environmental ethics have been about much "higher" organisms. It is certainly possible to argue that even microbial life is deserving of moral status, however it is also possible to overdo such claims. A position I call *Mariomania* has cropped up that is especially extreme in this regard. Mariomania is not yet a position espoused by professional ethicists and thus its nature and motivations are not discussed in the literature as deeply as a philosopher might wish. The basic idea seems to be that Martian life, no matter how

[1]Our initial reaction might also be to value Martian microbes highly because they are very rare—the only other form of life beyond Earth. However, finding life on Mars would actually be a good indication that life is ubiquitous in the universe.

K.C. Smith (✉)
Departments of Philosophy & Religion and Biological Sciences, Clemson University, Hardin Hall, Clemson, SC 29634, USA
e-mail: kcs@clemson.edu

J.S.J. Schwartz and T. Milligan (eds.), *The Ethics of Space Exploration*, Space and Society, DOI 10.1007/978-3-319-39827-3_14

lowly, should be accorded the highest possible moral standing—roughly on a par, not just with individual human beings, but with humanity collectively. As a consequence, it is claimed that human beings should forego any use they might wish to make of Mars, in perpetuity, whenever such efforts would conflict with the "interests" of Martian microbes. This contrasts with a more moderate position we might label *Mariophilia*, where it is allowed that Martian life would be extremely valuable and should be defended against *petty* demands of human beings, but also that human interests can in principle trump those of the Martians.[2]

It is possible that such claims are put forward as rhetorical devices and are not meant to be taken literally. At the very least, however, there are some influential thinkers who speak this way, with no apparent irony or exaggeration. Mariomania probably traces its origins to Carl Sagan's (in)famous quotation:

> If there is life on Mars, I believe we should do nothing with Mars. Mars then belongs to the Martians, even if the Martians are only microbes. The existence of an independent biology on a nearby planet is a treasure beyond assessing, and the preservation of that life must, I think, supersede any other possible use of Mars. (Sagan 1985, 130)

This is all Sagan says, so it is very unclear precisely what he means or why he feels as he does. Nevertheless, the idea seems to have taken root with a small but influential group of scientists. With each pronouncement, we find clues to their reasoning. For example, York says something very similar to Sagan:

> We must consider whether landing spacecraft on other planets is justified if those craft may carry with them the potential to destroy endemic ecosystems. Even if these ecosystems are composed of only microbial life, our ethical obligations are not thereby diminished. Life of independent origin of that of Earth must surely deserve valuation equal to that of our world. (York 2005, 73).

Here it seems to be that the independent origin of Martian life somehow motivates their being considered on a moral par with the entire terrestrial biosphere collectively.[3] Chris McKay, one of NASA's leading experts on Mars, goes even further to claim that we have, not just negative duties to refrain from interference, but positive ones as well:

> ...Martian life has rights. It has the right to continue its existence even if its extinction would benefit the biota of Earth. Furthermore, its right confer upon us the obligation to assist it in obtaining global diversity and stability. (McKay 1990, 195)

McKay turns the traditional question of whether humans should terraform Mars on its head, arguing that we should indeed transform Mars into a more hospitable place—but *for the microbes* rather than human beings! So Mariomania does exert a strong and perhaps growing influence on the astrobiological community. Thus far, it has largely gone unchallenged, in part because most professional ethicists are

---

[2]The terms *Mariophile* and *Mariomaniac* are meant as blanket terms covering all those who share these broad conclusions, though they might do so in a variety of ways and for a variety of reasons.

[3]York goes on to argue that, even if Mars is lifeless, an Aldo Leopold style "land ethic" may still be morally required.

unaware of it and in part because, at present, it has no practical import. It is time, however, to fill this critical lacuna with a careful analysis of Mariomania's justification and implications.

It's worth a word at the outset about what Mariomania is *not*. It can not simply be the claim that, *at present*, we should leave Martian microbes alone. At the moment, it is quite difficult for us to interfere with the Martian ecosystem in any systematic fashion and, even if we could, there is no compelling reason to do so. If all the Mariomaniacs are saying is that we should not do an impossible thing that nobody wants to do, theirs is a trivial claim. I give them more credit than this, assuming that their claim is rather that Martian microbes have a moral status that (1) will endure changing circumstance until we are actually able to exploit them for our own purposes and (2) would be sufficiently compelling at that point to block our taking certain actions even if they are in the interests of humanity.

14.2 The Role of Intrinsic Value

It is difficult to see how such conclusions could be defended solely on the basis of human interests (what philosophers often label *merely instrumental* value). For example, Martian microbes would undoubtedly have enormous scientific value to human beings and this has real moral weight. But such value can not provide an enduring obstacle to the pursuit of other human interests, since it will degrade over time. Thus, if the basis for the value of Martian microbes is ultimately scientific, then, once we have studied them for many decades and the prospect of major new discoveries begins to fade, so too will their moral value. So it seems that, if we truly have a robust duty to "leave Mars to the Martians" indefinitely, it must have a non-instrumental basis.

A common solution to this sort of problem is to place the favored entity on an ethical pedestal, as it were. As long as Martian microbes are simply one of many entities competing for ethical notice, it is only a matter of time before they lose out to some combination of other interests. On the other hand, if their needs are superior *in principle* to the instrumental considerations of others, then they are insulated from such threats. In ethical theory, the concept of *intrinsic value* is perfectly suited for this purpose; particularly as expressed by its most famous advocate, Immanuel Kant. Kant was greatly concerned with ethical theories like utilitarianism that seemed unable to defend human rights in a strong sense. To a utilitarian, a "right" is merely the temporary conjunction of particular conditions and consequently can disappear when circumstances change. Thus, nothing is truly sacred or enduring and everything is subject to the push and shove of the moral marketplace, where the interests of all entities compete in an ethical free-for-all. For Kant, any theory that allows such negotiations concerning basic rights does not deserve to be called moral—as he famously observed, "a moral being possesses dignity, not a price" (Kant 2012, 108).

Kant's solution was to adopt a strong conception of intrinsic value. If we can argue that human beings, by virtue of their intrinsic moral value, are *different in kind* from other beings, then we elevate them above the ethical fray. Kant conceives of intrinsic value as flowing from our uniquely human ability to reason.[4] Rational creatures have a moral value that is intrinsic to the kinds of beings they are, independent of any external factors. Their rights are thus truly enduring in that they are entirely immune from the effects of changing circumstance or opinion. As such, intrinsic value is the primary moral consideration and far more important than instrumental factors. All possessors of intrinsic value are morally equal to one another and morally superior to those without intrinsic value. To put it succinctly, morality is first and foremost about the interests of entities with intrinsic value.

Unfortunately, the justification for Mariomania is not entirely clear. However, both Cockell (2004) and McKay (1990) endorse the idea that microbes have intrinsic value. Cockell even asserts that the intrinsic value of *terrestrial* microbes is on a par with human value,[5] though in a way that suggests there might be a relationship to their instrumental value as well:

> Microbes have intrinsic worth equal to, if not greater than, that of any other species. They have instrumental worth so great that they are vital to our survival and the survival of all other multicellular organisms on Earth. (Cockell 2004, 149)

Even without such an explicit claim, it's hard to avoid the conclusion that something like intrinsic value is motivating Mariomania. Intrinsic value is an implicit aspect of the way many people think about moral value, in large part because it has been an element of classical moral theory for a very long time indeed. It is also perfectly positioned to solve the problem the Mariomaniacs face, by making a non-negotiable moral distinction that insulates its possessors, in principle, from the demands of others. The intrinsic value approach thus holds out a tantalizing hope: if Martian microbes truly have intrinsic value on a par with humans, then they clearly have rights comparable to humans that deserve the strongest possible protection. And this status is established in one fell swoop, without the need for complex discussions of practical tradeoffs that other types of arguments would require.

Kantian ethics is characterized by ethical rules that admit of no exceptions under any circumstances, since to allow an exception is to suggest that external conditions are relevant and thus invite the horrors of the moral marketplace. If killing another human being were only wrong in some situations, then skilled lawyers would find ways to argue that murder is perfectly reasonable in some circumstances. While inviolate moral rules sound laudable at first blush, in practice they are extremely

---

[4]It's important to separate two distinct claims here: (1) that some beings have a special, intrinsic, moral value and (2) that human beings are the only ones on Earth in this category. The first is a theoretical claim while the second is empirical. In particular, it is quite unfair to accuse Kant of unprincipled anthropomorphism, as he would have welcomed other beings to the community of intrinsic value had he been convinced they could reason in the requisite fashion.

[5]Cockell (2011) has distanced himself from full-blown Mariomania in more recent work.

difficult to apply realistically. One way to put it is that Kant places an effectively infinite value on all intrinsically valuable entities in the moral marketplace, which prevents the kind of haggling that can dilute our moral principles to meaninglessness. But it also prevents something markets do quite well: balancing competing interests in a systematic fashion.

In modern debates, the same difficulty is seen in PETA's famous adage, "a rat is a pig is a dog is a boy." While this makes a very strong appeal for the moral considerability of rats, by the exact same token it complicates any attempt to apply it in real life. It's one thing to grant that a rat has real moral standing, but it stretches moral credulity past the breaking point to suggest that, should a rat and a boy find themselves in a situation where one of them must die to save the other, we should flip a coin. So if we say that Martian microbes have intrinsic value in a Kantian sense, we immediately guarantee them protection so powerful it can block *any* action that might infringe on their interests. The result is an inability to apply our principles in a nuanced way when the interests of those with intrinsic value conflict, as they typically do in situations where we most need ethical guidance.[6]

14.3 The Prime Directive

The *Prime Directive* from *Star Trek* provides an excellent and especially relevant example of how an uncompromising ethical rule can create more problems than it solves. As every trekkie knows, Federation personnel in the *Star Trek* universe are bound, first and foremost, by Starfleet General Order #1 (colloquially referred to as The Prime Directive):

> As the right of each sentient species to live in accordance with its normal cultural evolution is considered sacred, no Star Fleet personnel may interfere with the normal and healthy development of alien life and culture. Such interference includes introducing superior knowledge, strength, or technology to a world whose society is incapable of handling such advantages wisely. Star Fleet personnel may not violate this Prime Directive, even to save their lives and/or their ship, unless they are acting to right an earlier violation or an accidental contamination of said culture. This directive takes precedence over any and all other considerations, and carries with it the highest moral obligation. (Grey et al. 2000)

The Prime Directive seems to be based on the idea that the ability of each culture to evolve independently is an overarching good such that other considerations, whether concerning the welfare of aliens or the Federation, should never be allowed

[6]It should be noted that the problem is not with intrinsic value per se, but rather with the *strength* of this value. It's certainly possible to defend a weaker notion of intrinsic value such that entities with intrinsic value are not necessarily more valuable than those without it. This restores our ability to negotiate conflicts between interests, but only by reintroducing the problem of the moral marketplace, since saying Martian microbes are intrinsically valuable in this weaker sense does not tell us much about how they should be treated relative to other beings.

to override it. This kind of thinking is also common in modern terrestrial politics, as the United Nation's Friendly Relations Declaration makes clear[7]:

> No State or group of States has the right to intervene, directly or indirectly, for any reason whatever, in the internal or external affairs of any other State. Consequently, armed intervention and all other forms of interference or attempted threats against the personality of the State or against its political, economic and cultural elements, are in violation of international law (United Nations 1970).

One of the interesting things about widespread cultural norms like these is that we are all exposed to them at an early age, before our critical abilities are fully formed. As a consequence, they tend to take root in our psyches and deeply influence our thinking, but in ways we don't always consciously appreciate. The idea that we should let other cultures "do their own thing," is thus a (largely) unexamined background assumption many bring to the Martian debates, particularly the science fiction buffs who dominate the space sciences. That's not to say that non-intervention is a bad moral principle, of course. But endorsing such a principle in a general (and typically vague) way and demanding its uncompromising application in every situation are very different things. A truly inviolate rule will often force us to choose between two unpalatable choices: blindly following it despite misgivings about the morality of its consequences in a particular case or making exceptions in some circumstances that undermine the very reason for its existence in the first place.[8] This is a point we often don't appreciate when thinking about ethical issues in the abstract, since the tradeoffs only become clear when we try to apply the principles so generated to real world situations.

There are currently no real world applications of Mariomania, but the use of the Prime Directive within the *Star Trek* universe provides a useful thought experiment. The members of Star Fleet see obedience to the Prime Directive as a sacred duty. As Captain Kirk puts it:

> A starship captain's most solemn oath is that he will give his life, even his entire crew, rather than violate the Prime Directive. (The Omega Glory 1968)

Thus, the Directive is followed not simply because it's a general order, but because it reflects the most deeply help ethical beliefs of the United Federation of Planets. In addition, it is seen as inviolate, admitting of no exceptions under any circumstances. In the words of Lieutenant Worf:

---

[7]It is perhaps worth noting the obvious here—that both the UN and Federation are clearly talking about our duties towards other *humans*, or at least other creatures with abilities much like humans. So even if we wholeheartedly endorse these principles in this context, it certainly does not follow that they apply to microbes.

[8]Indeed, the necessity of these kinds of moves can cause great confusion about the nature of morality. Thus, students in their first ethics classes will often say things like, "Well, this is not the moral thing to do, but it's the right thing," suggesting that the right thing to do is often different from our moral principles. Obviously, this sort of confusion can greatly undermine our faith in the possibility of an objective morality—the defense of which is, at least historically, the point of inviolate principles in the first place.

The Prime Directive is not a matter of degree. It is an absolute. (Pen Pals 1989)

Since the Prime Directive expresses the most deeply felt ethical intuition of the Federation in the form of an inviolate rule, one might expect its application to be relatively straightforward. But in fact the clash between the ideals of the prime directive and practical reality is a recurrent theme in the *Star Trek* universe—to the point where it could be the basis of a drinking game.

On the one hand, we have examples of Federation personnel following the directive even when it seems extremely unlikely that doing so is ethical. Consider an episode of *Star Trek: Enterprise* (Dear Doctor 2002) where the Starship Enterprise discovers a primitive culture being decimated by a massive pandemic. The aliens seek help, and the crew investigates. They quickly discover that there are two distinct subspecies on the planet, with the disease affecting only one of them. The ship's doctor projects that the vulnerable subspecies will be extinct within 200 years, and develops a medication that can eradicate the disease. However, the captain decides to withhold the treatment in deference to the Prime Directive. Whatever we might think about the ideals behind the Prime Directive in general, this seems to be a case where non-intervention is almost certainly immoral. Condemning billions of beings to a fatal illness, not to mention allowing the extinction of an entire subspecies, is very, very bad—particularly since it could so easily be avoided. Almost everyone would allow that some degree of flexibility in applying the directive would lead to a morally superior outcome in this situation.[9]

On the other hand, *Star Trek* characters routinely make exceptions to the Prime Directive when it suits their purposes. Captain Kirk famously believed, not only that the Prime Directive applied only to "living, growing civilizations," but that he was well positioned to make the decision as to which civilizations were worthy of protection. Predictably, this results in highly questionable exceptions to the rule. Consider an episode of the original series (The Apple 1967) where the Enterprise encounters an alien culture that appears truly idyllic. The people are friendly, happy, and peaceful—to the point where they lack even the *concept* of murder. But when the computer that manages this paradise attempts to destroy the Enterprise, Kirk decides the culture is stagnant and destroys the computer instead, along with the culture it made possible. Despite explicit claims that the Prime Directive is above such considerations, all too often its application comes down to the strategic importance of the planet in question or the welfare of the ship and crew.

In part, of course, these conflicts exist because they provide for interesting plot twists. However, they also represent genuine moral dilemmas produced by the conflicting interests of entities with clear moral value. As such, they demand complex and nuanced discussions that quickly exhaust the moral resources afforded by a one size fits all rule. It seems very likely that the case will be no different for Martian microbes: even if we allow that they have real (intrinsic or otherwise)

[9]To be sure, there are other problems with the Prime Directive. For example, it often seems to be little more than an instance of the naturalistic fallacy. In this particular case, it has also been argued that it invokes a particularly unsavory form of social Darwinism (Dvorsky 2007).

moral value, this can not mean they will *always* win out when their interests conflict with those of humanity. It is thus not defensible to make the kind of blanket statements the Mariomaniacs do.

14.4 Reception of Mariomaniacal Ideas

In my conversations with other ethicists, I often encounter incredulity concerning Mariomaniacal claims. "Surely," I hear, "these people do not mean to imply that *no* human interest in Mars could *ever* be sufficiently compelling that it should be pursued, merely because they might harm microbes!" So I decided to collect some rough data on what people actually think of these kinds of claims. In March of 2015, I created an online survey and advertised it on two prominent philosophy blogs (*PEA Soup* and *Daily Nous*) where ethical issues are often discussed. The survey received a total of 181 responses, 103 from professional ethicists and 78 from interested laypeople. The survey asked respondents to rate their initial take as to the plausibility of two key moral claims, one by Charles Cockell and another by Carl Sagan.

Figure 14.1 shows the response to Cockell's claim that microbes have intrinsic moral value on a par with all other species (including humans). This particular quotation actually refers to terrestrial microbes, but I wanted to gauge people's reaction to an explicit claim of strong intrinsic value applied to microbes. The reactions of the two groups of respondents were similar, though professional ethicists were slightly more skeptical about it than lay people. Generally speaking, the claim is not considered credible, with only about 1 in 14 people thinking it is probably or patently true and only about 1 in 4 finding it even a plausible

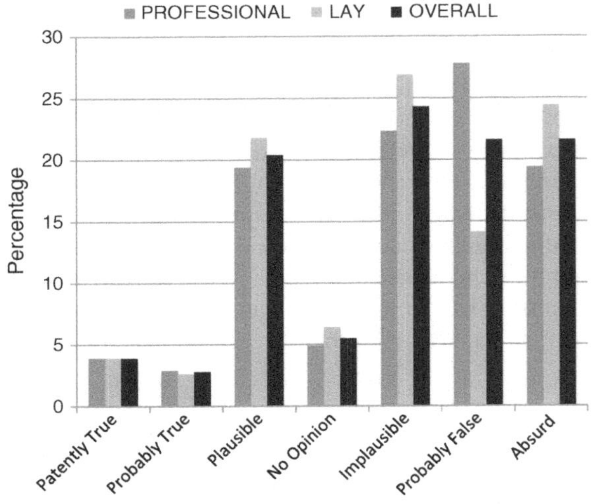

Fig. 14.1 Initial reaction to the claim, "Microbes have intrinsic worth equal to, if not greater than, that of any other species" (Cockell 2004) by lay and professional ethicists

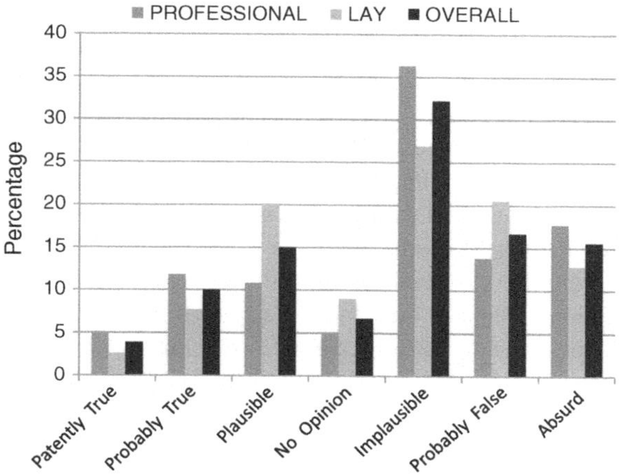

Fig. 14.2 Initial reaction to the claim, "If there is life on Mars, I believe we should do nothing with Mars. Mars then belongs to the Martians, even if the Martians are only microbes" (Sagan 1985) by lay and professional ethicists

possibility. By contrast, nearly half of the professional ethicists considered it either probably false or absurd.

Figure 14.2 shows the response to Carl Sagan's famous claim that founded Mariomania. Here again, the reception was similar between the two groups, though professional ethicists were a bit more sympathetic to it than lay people. The reaction was still largely negative, with only about 1 in 6 of the professional ethicists and about 1 in 10 of the lay people feeling it was probably or patently true, and about a third of all respondents finding it either probably false or absurd.

So the results are mixed. On the one hand, Mariomaniac claims are clearly not popular, with only about 4 % of respondents clearly identifying themselves as within the Mariomaniac camp and about 70 % of respondents viewing these sorts of claims as dubious. On the other hand, approximately 30 % of respondents seem willing to consider at least some elements of Mariomania. I do not mean to suggest that popularity is always a good guide to the truth, of course. However, the lack of support for these ideas, especially within the ethics community, should at least place the burden of proof squarely on Mariomaniacal shoulders. If their position is to be taken seriously, Mariomaniacs will have to provide the theoretical arguments that can turn back such negative intuitive reactions.

14.5 A Counterargument

Mariomania has been sheltered from direct critique in part because few professional ethicists are aware of it, but also because it has no practical implications at the moment. Thus, while it is true that a Mariomaniac would urge caution concerning how we handle the initial investigation of Martian microbes, so too would a great many people with more moderate sentiments, and thus there is no immediate conflict between Mariomania and Mariophilia. Yet I suspect many would be very concerned about Mariomania if it were ever to become widely accepted. Indeed, such concerns have already been expressed in the context of Martian exploratory missions (Battaglia 2015).[10]

We must therefore take care to distinguish between the true Mariomaniac, who defends the interests of Martian microbes in enduring and uncompromising terms, and the Mariophile, who also believes Martian microbes need to be protected, but in a more qualified fashion. To some extent, this is a difference of degree, and thus full-blown Mariomania is the endpoint of a spectrum of possible positions. This ambiguity, combined with the enormous variety of possible positions and justifications, makes it hard to give a specific face to *the* alternative to Mariomania. But it is possible to present a rough schema of an argument that might be offered in opposition to Mariomania. In order to do this, however, we must make some initial assumptions:

1. There are microbes on Mars (whether of independent origin or not) but no other kinds of Martian life.
2. We have studied this life extensively and answered most of the major scientific questions about it.
3. There is some opportunity on Mars that, if pursued, would be of significant benefit—not just to individual humans, but also to humanity more generally.
4. It is feasible to pursue this opportunity, but not without a major negative impact on the Martian microbes.

To take the classic (if futuristic) example, we might suppose that humanity will one day be able to create something like a second Earth by terraforming Mars, but doing so would also destroy the natural habitats of the indigenous microbes. Given this setup, consider the following argument:

Premise Humans beings are in the highest known category of moral value (it matters not for present purposes whether they are alone in this regard or which ethical theory is used to justify this status). We thus have an unambiguous obligation to serve human interests, particularly the interests of large numbers of human beings or of humanity collectively.

Premise The claim that Martian microbes are of *equal* moral value to humans, though possible, seems unlikely.

Therefore To the extent that compromise is not possible and we are forced to choose whose interests to maximize, we are *obliged* to err on the side of humans.

[10]This recent posting by an editor of *Scientific American* on their blog expresses concern about restrictions on scientific investigations of Mars based on an overly strong desire to prevent forward contamination of Martian microbes. The co-Chairs of NASA's planetary protection commission posted a defense of their policies which draw, not on Mariomaniac arguments, but rather on purely pragmatic considerations as to how best to do the science. This is a crucial distinction, because there is much more consensus on both the *scientific* value of Martian life and the best way to conduct the science (at least in the short term) than on more abstract considerations of moral value.

One key difference between this argument and the apparent thinking of the Mariomaniacs is an appreciation of the *opportunity costs* should human beings forego the exploitation of Mars and its resources. Part of what drives our non-interventionist intuitions is probably the implicit notion that the status quo does not require explicit moral justification. Thus one might think, "Humans have done just fine so far without using Mars, so it's best not to start doing things differently if that might present new problems." But when the matter under debate is whether we should take advantage of an opportunity to *improve* the status quo in morally relevant ways, this attitude is highly problematic. If the debate over whether and how to exploit Mars eventually becomes real, it will be because we have found a way to use some aspect of Mars to serve human interests. This use will mean making at least some humans better off, which will bring up complex moral tradeoffs. It's one thing to say that we should leave Mars alone when this can be done without cost, but quite another to say we should continue to do so when this requires human beings to lose out on significant moral goods.

14.6 Objections and Replies

Any thoughtful commentator will have to allow, at the very minimum, that the interests of Martian microbes and humans are entwined to some extent (if not as closely as those of humans and terrestrial microbes). If nothing else, the scientific value of Martian life clearly justifies a conservative approach to our early interactions with them, whatever their status may be in terms of independent moral value. On the other hand, if this is all there is to the moral value of microbes, then our commitment to protecting the microbes will evolve over time and we may ultimately be *obliged* to entertain proposals that would harm them in various ways. For example, is providing a 10 % boost to the standard of living on Earth sufficient reason to wipe out Martian microbes forever? Is preventing the deaths of 1000 human beings? These questions are both perfectly legitimate and quite difficult to answer in an a priori fashion.

It's also critical to understand that rejection of Mariomania does not necessarily imply acceptance of problematic alternatives. One idea might be that granting the right of humans to pursue their interests at the expense of other beings will inevitably lead down a slippery slope. But even if we believe, should push truly come to shove, that human interests *always* trump microbial ones, this need not imply that *individual* humans or *particular groups* of humans should always get their way. Even if they have no intrinsic value, Martian microbes have sufficient instrumental value (including the fact that many humans derive pleasure from contemplating an unspoiled Mars) to override any number of petty human interests (Smith 2014). So while concerns over the slipperiness of the slope are well taken, they certainly don't provide sufficient reason to adopt full-blown Mariomania.

Another potential move is to accuse those who argue in favor of exploitation of space resources (microbial or otherwise) of *anthropocentrism*. Of course, any time we assert human interests over those of other entities we could be said to be "anthropocentric" in *some* sense. But it matters greatly *why* we are favoring humans. In philosophical circles, anthropocentrism typically refers to a particularly simplistic conception of moral value such that humans are considered more valuable than other entities simply because they are (*biologically*) human (Manson 2012). This argument is indeed problematic, but it's hardly fair to suggest that it captures the motivation of all those who oppose Mariomania. Almost all ethicists believe there are crucial properties, the possession of which heavily influences moral value. We can debate what those properties are and who possesses them to what extent, but there is nothing wrong with line drawing per se just as there is nothing wrong, in principle, with finding humans to be morally special. Humans do, after all, seem to possess several morally relevant traits either uniquely or to a unique extent, at least among life on Earth. Of course, whether this is actually true is still an open empirical question in the terrestrial context and even more so in the context of astrobiology. It is worth nothing that it is even possible for an opponent of Mariomania to believe that some forms of extraterrestrial life could be *more* valuable than humans—for example, by virtue of possessing superior rational capacity. Labeling such a position "anthropocentric" would be deeply misguided.

If anything, it's the *Mariomaniac* who is in danger of overly simplistic moral reasoning. It is extremely unclear just what it is about Martian microbes that supposedly grants them privileged moral status, especially since they are unlikely to have any of the properties traditionally associated with such status (e.g., sentience, sociality, reason).[11] If the only reason they are considered morally special is because they are *Martian*, then the Mariomaniac is guilty of a *Mariocentrism* every bit as pernicious as the anthropocentrism of old.

Finally, one might object that we could be wrong about some of the key claims in the argument schema—for example, the fact that humans tend to *think* we are morally special is not good evidence that we actually *are*. This is certainly true, but it doesn't favor one side of the debate over the other, as either side could be wrong. The best we can ever do is to examine the options as objectively as we are able and make a decision based on the understanding this produces. The claim that humans are morally more valuable than Martian microbes is not a very controversial one as ethical claims go and it thus seems entirely reasonable to adopt this as our working assumption until and unless we find compelling reasons to abandon it. In a zero sum conflict of interest of the sort envisioned in the terraforming case, we are forced to choose and, while we *might* be acting immorally towards the microbes by favoring humans, it's much more *likely* that we would be acting immorally towards humans by favoring the microbes.

---

[11]They are alive, though since this is a property shared with all life on Earth, it doesn't help much in deciding *relative* ethical value. It might also be argued that they have the *potential* to develop these other ethically salient properties and thus deserve moral privilege on this basis. It is certainly possible to argue this way, though again such potential is presumably quite common. Moreover, plausibility is strained considerably when we realize that the timeframe needed for evolution to realize this potential is probably on the order of several *billion* years.

14.7 Conclusion

> Burning bridges behind you is understandable. It's the bridges before us
>
> that we burn, not realizing we may need to cross, that brings regret.
>
> —Anthony Liccione

The idea that Martian microbes have ethical value deserving of serious consideration is a perfectly legitimate one. However, the Mariomaniac view that this value is so great that humanity should forego any opportunity whatsoever arising on Mars if it conflicts with microbial interests is not credible for both theoretical and practical reasons. Theoretically, Mariomania seems to be driven by the idea that Martian microbes would possess strong intrinsic moral value. Yet those who defend this view provide very little argument for such a conclusion and it is deemed implausible by a clear majority of professional ethicists. Practically, such a view tends to produce inviolate ethical rules that don't allow for nuanced discussion of the complex tradeoffs most controversies in ethics present. The Prime Directive provides a fictional example of just such a rule and the problems it creates. Federation crews are constantly forced to choose between (1) obeying the Directive even when its consequences seem immoral and (2) making ad hoc exceptions which undercut the moral principle behind the rule. Should Mariomania ever become the dominant view, we will eventually face much the same situation with respect to Mars: in particular cases, we will waffle between allowing human suffering by forgoing any significant changes to Mars and finding clever ways to rationalize human-serving actions inconsistent with our avowed Mariomaniacal beliefs.

In response, I offer a counterargument to Mariomania that probably mirrors the thinking of the vast majority of professional ethicists, notwithstanding considerable disagreement over details. We must explicitly acknowledge the obvious: that while it is clear (at least to the extent that anything in ethics is clear) that humans have very high moral value, it is much less clear if the value of microbes is anywhere near as great. Since any decision not to exploit Martian resources will represent opportunity costs with morally significant impacts on humans, we should err on the side of favoring humans. At the very least, we must admit that it is possible, in principle, for human interests to overbalance those of Martian microbes in some circumstances.

There is a company that produces amusingly ambivalent bumper stickers, one of which pokes fun at bumper sticker environmental slogans:

> The benefits of environmental protection measures should be thoughtfully weighed against their costs, and the sound ones enacted.

This is a joke, of course, but one that also makes an important point. It is all too easy to espouse an ethical principle whose intuitive appeal hides the fact that is it unclear, impractical or even contradictory. Mariomania is such a slogan. While it appeals to noble intuitions, it does so in a way that can not be taken literally.

Indeed, by taking the place of more realistic discussions of our ethical obligations, it could actually *impede* moral progress. It is far more honest, if less satisfying, to be upfront about the complexities involved in these situations. If asked whether we should exploit some environmental resource, whether it's in the Antarctic wilderness on Earth or a Martian environment rich with indigenous microbes, the only answer both succinct and honest is, "That depends."

References

Battaglia, M. (2015). Protecting' Mars from contamination a half-baked idea? *Scientific American Blog*. November 6, 2015. Accessed online at http://blogs.scientificamerican.com/observations/is-protecting-mars-from-contamination-a-half-baked-idea/ on 11/28/15.

Cockell, C. S. (2004). The rights of microbes. *Interdisciplinary Science Reviews, 29*, 141–150.

Cockell, C. S. (2011). Ethics and extraterrestrial life. In U. Landfester, et. al. (Eds.), *Humans in outer space: Interdisciplinary perspectives* (pp. 80–101). New York: Springer.

"Dear Doctor." *Star trek: Enterprise*. UPN. January 23, 2002. Television.

Dvorsky, G. (2007). Star trek's prime directive is stupid. Accessed online http://www.sentientdevelopments.com/2007/01/star-treks-prime-directive-is-stupid_13.html on 11/28/15.

Gibson, E., McKay, D., Thomas-Keptra, K., & Romanek, C. (1997). The case for relic life on Mars. *Scientific American, 277*, 58–65.

Grey, C., & Chris S. J. (2000). Starfleet general orders. Accessed online at http://www.st-minutiae.com/articles/treaties/general_orders.html on 11/28/15.

Kant, I. (2012). *Groundwork of the metaphysics of morals* (2nd Edn., M. Gregor & J, Timmermann, Trans.). London: Cambridge University Press.

Manson, N. (2012). Anthropocentrism, exoplanets and the cosmic perspective. *Environmental Ethics, 34*(3), 275–90.

McKay, C. P. (1990). Does Mars have rights? An approach to the environmental ethics of planetary engineering. In D. MacNiven (Ed.), *Moral expertise* (pp. 184–197). New York: Routledge.

"Pen Pals." *Star Trek: The next generation*. In syndication. May 1, 1989. Television.

Sagan, C. (1985). *Cosmos*. New York: Ballentine Books.

Smith, K. C. (2014). Manifest complexity: A foundational ethic for astrobiology? *Space Policy, 30*, 209–214.

"The Apple." *Star trek*. NBC. October 13, 1967. Television.

"The Omega Glory." *Star trek*. NBC. March 1, 1968. Television.

United Nations General Assembly. (1970). *Declaration on principles of international law concerning friendly relations and cooperation among states in accordance with the charter of the united nations*. New York. October 24, 1970.

York, R. (2005). Toward a Martian land ethic. *Human Ecology Review, 12*, 72–73.

Chapter 15
The Aesthetic Objection to Terraforming Mars

Sean McMahon

15.1 Introduction

Mars is a remote planet, removed from us by millions of miles and still further obscured by centuries of misunderstanding and fantastical speculation. But with every new measurement and photograph gathered by our telescopes, rovers and satellites, Mars emerges a little further from the fog. Despite its barrenness, the new world shifting into focus is starkly, grandly beautiful. In this chapter I discuss the challenge that this beauty poses to the case for *terraforming* Mars—that is, forcing the planet into an artificially Earth-like, humanly habitable state by warming the surface to release frozen water and carbon dioxide, which will trap more heat, thicken the atmosphere and flood the lowlands; and seeding life forms to oxygenate the air. For now, this old science-fiction scenario is merely an entertaining thought-experiment for planetary scientists and environmental ethicists, but putting it into effect it may become technically and economically achievable in coming centuries.

Meanwhile, we have plenty of time for debate. The ethical controversy pivots not on whether we ought to begin terraforming Mars right away but on whether doing so once the necessary technological and economic progress has been made will be morally permissible or even recommended. This debate has recently been reviewed by Schwartz (2013), whose "*prima facie* assessment of terraforming is that it is morally recommended insofar as it would contribute to our environmental education" (on the assumption that Mars is devoid of indigenous life). In his view, terraforming Mars would be a kind of planetary-scale research project, likely to improve our understanding of ecology and biogeochemistry and therefore to help us meet our moral duties in the stewardship of the Earth. Schwartz is unconvinced by objections based on appeals to the value of existing Martian environments, arguing that no compelling case has been made that Mars in its current state is more

S. McMahon (✉)
Department of Geology and Geophysics, Yale University, New Haven, USA
e-mail: sean.mcmahon@yale.edu

J.S.J. Schwartz and T. Milligan (eds.), *The Ethics of Space Exploration*,
Space and Society, DOI 10.1007/978-3-319-39827-3_15

valuable than the scientific and educational dividends of terraforming. However, he acknowledges that Mars exploration is at an early stage and that caution about terraforming is advisable until certain questions about the nature and value of Mars are answered. We do not yet know whether Mars is inhabited or the extent of Mars' scientific or aesthetic value.

Although I broadly agree with this agnostic conclusion, in this chapter I attempt to show what we *do* know about Mars' aesthetic value. I argue that the known aesthetic qualities of Mars already furnish a serious objection to terraforming, and that this objection may strengthen as human aesthetic engagement with Mars becomes more immersive. I begin by discussing the ways in which other authors have dealt with aesthetic objections to terraforming. The most effective response, I think, is the claim that terraforming would make Mars more rather than less beautiful. I contextualize this claim by surveying the history of ideas about real and imagined Martian landscapes, drawing attention to the way that Mars has repeatedly fallen short of our geocentric expectations. After sketching my own impressions of the beauty of Mars, I then highlight the use of geocentric distortions for aesthetic effect in popular contemporary images of Mars, which might be taken as evidence that terraforming Mars would indeed enhance rather than diminish its aesthetic qualities. In response, I suggest that Mars may offer distinctively Martian forms of beauty yet to be generally recognized or appreciated, partly because of a general lack of serious aesthetic engagement with the in situ photographs already available. I conclude that Martian aesthetic qualities provide a clear if defeasible objection to terraforming the surface of Mars, while it remains to be seen how valuable these qualities are felt to be once they are better known. I am not arguing that terraforming is morally prohibited in the final analysis. Rather, I am attempting to flesh out a serious objection that has not hitherto been explored in depth and to give it a fuller and more considered hearing. This objection may not ultimately be decisive, but it should at the very least be weighed carefully together with other considerations in the debate about terraforming.

15.2 The Aesthetic Objection to Terraforming Mars

Up to a certain point, Mars can be explored, exploited and even colonised in ways that control and minimise aesthetic damage. The "planetary park" system proposed by Cockell and Horneck (2004) would afford areas of great natural beauty the protection of legal instruments analogous to those governing national parks and other protected areas on Earth. These laws and treaties make explicit and legally binding the acknowledged moral duty of preserving natural beauty. The Antarctic Treaty, covering a vast continent whose wonders few of us will ever enjoy first-hand, mandates the protection of "the intrinsic value of Antarctica, including its wilderness and aesthetic values" (Protocol on Environmental Protection Article 3, paragraph 1). Work in the treaty area is required to avoid "degradation of, or substantial risk to, areas of biological, scientific, historic, aesthetic or wilderness significance" [ibid. Article 3, paragraph 2(b)(vi)]. Similar requirements would obtain within Cockell and Horneck's Martian parks, outside of which development could proceed in accordance with other priorities.

But such careful spatial discrimination would not be possible if Mars were to be terraformed; raising the global temperature would unavoidably flood the vast Martian lowlands, and the intensification of the water cycle would cause massive erosion in the highlands. Terraforming Mars would radically transform its landscapes and irreversibly change their aesthetic qualities. Many authors have supposed that this transformation would be for the worse—a destruction of beauty on a planetary scale. For this reason, aesthetic considerations are often counted among the reasons not to terraform Mars; Robert Sparrow, for example (1999), claims that to do so would demonstrate the vice of "aesthetic insensitivity", a "serious defect of moral character".

Even so, and despite the fact that aesthetic considerations have long played important roles in both justifying and motivating campaigns for the conservation of natural environments on Earth, the preservation of beauty might seem rather trivial compared to other considerations in the debate about terraforming. On the one hand, Mars may host an indigenous biosphere, which according to some writers would have a moral status of its own that demands our unconditional respect. On the other hand, it is commonly claimed or implied that terraforming Mars may be necessary for the survival of the human race, and that all objections pale into insignificance beside this overwhelming imperative, which would arguably make terraforming not only permissible but morally recommended. Sparrow concedes that in this case terraforming would be morally permitted, although if we were virtuous we would undertake it regretfully, "fully aware of the beauty that [we would be] destroying" (ibid., 240). Schwartz (2013, 11) regards aesthetic objections as insignificant compared to the existential risk that terraforming Mars would supposedly mitigate: "threatened with humanity's demise, concerns of this sort fall away".

This kind of defence of terraforming is probably misguided. Clearly, few obligations trump the need to avert human extinction; most thought experiments in ethics can be swiftly concluded once it is stipulated that only one course of action leads to the preservation of the human race. But it is difficult to envisage a situation in which terraforming Mars would be the most practical way to avert extinction, let alone the only way. Human habitation can be accommodated on Mars without terraforming the planet. For human colonists, terraforming might be motivated by considerations of health, safety, ease and comfort, but it is hard to see how the survival of humanity itself could be at stake; if it were, it would be faster and easier to adapt human life to Mars than to adapt Mars for human life.[1] Admittedly,

[1]Perhaps Mars, if rapidly terraformed and maintained in a habitable but under-populated state, could represent an "emergency backup" planet. If some unavoidable cataclysm threatened at short (but not too short) notice to wipe out the population of Earth (but not Mars), it may be that only a terraformed Mars would have the carrying capacity to accommodate immediately and then to sustain billions of interplanetary refugees. If we are seriously concerned about such eventualities, terraforming is certainly motivated. But since no cataclysm appropriate to this highly contrived scenario is known to science, terraforming an entire planet seems an unreasonably extreme precaution.

terraforming Mars would increase its carrying capacity so that many more humans could live there, and for this reason might be favoured by those who advocate the unbounded expansion of the human race (so as to facilitate "the greatest happiness for the greatest number", for example). But this ambition is, to say the least, more open to question than the obligation to preserve human existence per se, and far less compelling as a reason for seeing terraforming as morally recommended.

Perhaps a more troubling reply to the aesthetic objection to terraforming is that Mars is not (or cannot be shown to be) *objectively* beautiful: "beauty is in the eye of the beholder". In Schwartz's (2013, 23) comment on Sparrow, he seizes on the difficulty of constructing a robust objectivist account of beauty. But in fact, there are alternative accounts of beauty under which Sparrow's charge of "aesthetic insensitivity" retains at least some of its force (Sparrow 1999, 2015). And more importantly, I think, *whatever* theory of beauty we adopt, once we admit to finding a thing beautiful ourselves, we acknowledge a value in it that gives us a *prima facie* reason for treating it with caution. Thus, Sparrow writes, "what matters is that Mars is beautiful [...] rather than the particular philosophical account we give of the nature of aesthetic judgements or properties" (2015, 174). Whoever is persuaded that Mars is beautiful will be justified in offering some proportionate resistance to the effacement of this beauty. The destruction of a beautiful thing is sometimes justified in the end by good reasons, but I agree with Sparrow that it should be a cause for regret nonetheless.

We are perfectly capable of discussing whether a thing is beautiful without needing to establish beforehand the metaphysical or epistemological status of aesthetic properties. Meta-aesthetical debates are arguably unnecessary in this context; here, what matters is the magnitude and significance of the aesthetic qualities that we actually perceive in Mars and how terraforming will change them. And it is here that the aesthetic objection to terraforming encounters what I think is probably its most serious obstacle: the plausibility of the suggestion that terraforming Mars would actually increase its aesthetic value. On Earth, beauty is found in the various hues, patterns, shapes, movements and sounds of animals, plants and other living things, the strong colours transmitted through our clear, bright skies, and the spectacle of water massed impressively into oceans, lakes and rivers. Because Mars lacks these delights but would gain them through terraforming, it is often suggested that aesthetic considerations, insofar as they carry any weight at all, actually motivate terraforming (e.g. McKay 1990). In his *Mars* trilogy of epic sci-fi novels, Kim Stanley Robinson embodies this view in the character of Sax Russell: "If [after terraforming] there are lakes, or forests, or glaciers, how does that diminish Mars's beauty? I don't think it does. I think it only enhances it" (2001, 174).

How are we to determine whether, on balance, aesthetic considerations militate for or against terraforming? Knowing as little as we do, it is difficult to weigh the value of the beauty that terraforming would destroy against the value of the beauty that terraforming would create. If we wish to gain some idea of the former, it is essential not only that we pay close attention to the sensible qualities of Mars, but that we pay attention in an open-minded, receptive, and scientifically well-informed way, alive to the aesthetic possibilities of a new and unfamiliar world.

15.3 The Search for Beauty on Mars

Since Galileo first resolved the reddish wandering star into a planetary disc, learned, lay and literary views of Mars have changed many times. Forgivably enough, in the absence of data its climate and habitability were generally viewed as earthlike. Ambiguous dark and light features were widely assumed to be continents and oceans, and William Herschel's observations of the Martian atmosphere prompted him to suggest that intelligent Martians "probably enjoy a situation in many respects similar to our ours" (1784, 273). Informed opinion began to turn against these assumptions in the late nineteenth century, although the Italian astronomer Giovanni Schiaparelli was not alone in continuing to insist that Mars was "not a desert of arid rocks. It *lives*" (1888, 3). Schiaparelli divided the Martian surface into regions and blithely named them Eden, Elysium, Arcadia and Utopia; between these he thought huge water-filled channels could be discerned. Notoriously, the American astronomer Percival Lowell mistranslated Schiaparelli's term *canali* (channels) as "canals", claimed to be able to see them through his telescope, and inferred that Mars was home to a technologically advanced alien civilization undertaking massive engineering works to distribute water on a slowly drying world. This idea found a receptive public and inspired H.G. Wells' *The War of the Worlds* (1898). Although the canals were soon debunked to the satisfaction of the scientific community, the possibility of humanly or superhumanly intelligent Martians—and the emotive idea that Mars is either a "dying" or a "dead" world—has retained a certain currency ever since. Well into the twentieth century, even in the most respectable scientific circles, it was thought that Mars was probably inhabited, its seasonally shifting colours created by deciduous vegetation.

Thus, Mars has always been the screen onto which human dreams and fantasies have been projected. The more we have learned about Mars, the more sorely our expectations have been frustrated. In 1964, NASA's Mariner 4 probe flew by Mars and imaged its barren, cratered surface. That Mars looked more like the moon than the Earth was to prove the first of many disappointments. When NASA technicians obtained the first colour photographs from the surface of Mars in 1976, they were unprepared for the intensity of red colouration in the sky. Unsure and "unwilling to commit [themselves] publicly to this provocative display," they therefore recalibrated the images before releasing them: the pink sky was toned down to a "neutral grey" that became bluish when the image was reproduced (Viking Lander Imaging Team 1978). The sight of a blue sky on Mars delighted the media, who were correspondingly disappointed when the original images were released a few days later, having been determined to be accurate (ibid.). On the internet, one can still find the protests of conspiracy theorists who never accepted this revision, and continue to accuse NASA of literally covering up the blue colour of the true

Martian sky; online videos repeating this falsehood have been watched hundreds of thousands of times.

Devoid of blue skies, greenery, and canals, Mars has consistently failed to live up to the unrealistic demands we have made of it.[2] Perhaps in an obscure way the proponents of terraforming are responding to this sense of failure. But, as Holmes Rolston has put it, "we ought not condemn Mars because it failed to be Earth" (1986, 172), especially if we are serious about evaluating the aesthetic consequences of terraforming. In order to discover what aesthetic qualities Mars may possess, we must instead become receptive to Mars as it actually is.

What, then, is Mars really like? In my own work as a planetary scientist, I have become relatively familiar with the available imagery of the Martian surface. Besides the picture-postcard panoramas usually chosen for press releases, there are now in the public domain thousands of geological snapshots captured by NASA's Spirit, Opportunity and Curiosity rovers, and again thousands of exquisitely detailed satellite photographs from the HiRISE ("High-Resolution Imaging Science Experiment") camera aboard Mars Reconnaisance Orbiter (MRO). To a geologist, the landscapes recall the Canadian High Arctic or the rocky deserts of Libya. The sky, however, is a strange inversion of the Earth's: brown, orange, lilac, yellow or olive green during the day but electric blue in the glow of the setting sun. Sometimes the blue returns before sunrise in high, feathery wisps on the brightening sky—noctilucent clouds in the upper atmosphere. The little moon Phobos rises and sets twice daily, passing in front of its smaller, slower twin, Deimos. In the warmer seasons, low-latitude hillsides and crater-flanks are painted in fine, dark, parallel strokes by trickles of saltwater. On the floors of craters, black and red sand and dust pile into ripples and dunes, the largest in the solar system. Dust devils are ubiquitous, scrawling dark curlicues on the desert floor like the tracks in a cloud chamber. Compared to Earth, Mars is stark and simple. Undisturbed by the messy influence of biology, Martian features have a mathematical regularity; the same patterns recur at many different scales, like those of a fractal: frost polygons, craters, ripples and dunes, fractures, veins and valley networks.

The aesthetic experience of standing on Mars or flying over it is no doubt very different from browsing NASA photographs from the comfort of an office chair. Moreover, the field sites chosen for landers and rovers are always rather featureless plains where the risk of a bumpy landing is lowest; the most classically sublime Martian landscapes have not been seen from the ground even by robots. Nevertheless, like many of those who have studied the photographs in detail, for me Mars has developed a complex emotional significance, something like a "sense of place" (although it seems strange to apply that concept on a planetary scale). In short, it seems to me that Mars possesses a great and distinctive beauty.

---

[2]I owe this thought to Tony Milligan.

15.4 Terraformed Images

I am not, of course, alone in finding the Mars revealed to us in recent years powerfully beautiful; indeed, the scientific team behind the HiRISE images distributes them to the public through a website called "Mars is Beautiful" (http://beautifulmars.tumblr.com). But it is fair to say that images like these have not yet permeated the public consciousness. Unlike Pluto, whose surface features were imaged for the first time this year, Mars carries a heavy load of cultural baggage, including preconceptions about its appearance, which scientific data have yet to displace fully. In particular, our sense of how Mars might be beautiful is heavily coloured—literally—both by earthly expectations and by crude stereotypes of the "red planet". Although the Martian surface has now been photographed tens of thousands of times, "artists' impressions" continue to propagate in news articles and other products of mass media. What is remarkable about these images is not that they are inaccurate but that they consistently deviate from reality in the same direction: towards a Romanticized, sublime and distinctly earthly vision of desolate rocky grandeur, albeit smothered with red tones. In what I call "terraformed images", the gentle, buttery colours of Mars are replaced with the hot reds of a sunset on Earth, with white-golden highlights and strong contrasts; instead of smooth pale skies we see dramatic cloud formations and thick banks of fog bristling with crepuscular rays. A Google image-search for "Mars beauty" (without "") or similar terms reveals dozens of these images, which appear both more frequently and with higher priority than real photographs in the results. Almost all the images purportedly of the Martian surface are artists' impressions incorporating features of Earth's natural beauty that cannot be found on Mars. Particularly popular are the computer-generated images by the Dutch artist Kees Veenenbos, which in 2006 were widely circulated with a false attribution to the MRO HiRISE instrument (not the artist's own doing). Veenenbos's renderings (which still feature with misleading prominence in the results of a Google image-search for "HiRISE") combine the real topography of the Martian surface with sharp, white, Earth-like lighting and atmospheric effects; the red-brown, light-scattering haze of Mars is gone. Whatever beauty these images possess, it is not the beauty of Mars.

Terraformed images thus embellish Martian landscapes with the earthly forms and palettes conventionally used to achieve aesthetic effects in terrestrial art. Their popularity compared with that of real photographs of Mars might seem to imply that Mars becomes beautiful only by borrowing the beauty of Earth, and therefore to lend some weight to the case for terraforming on aesthetic grounds. But I think posterity will draw a different lesson from these images. They remind me of the paintings of lions and elephants in medieval bestiaries, executed by monks who had never seen a true likeness of these animals and who singularly failed to capture their graceful physiognomies, giving them instead the faces of cats, dogs, or men. Popular images of Mars exhibit the same tendency crudely to substitute the familiar for the unknown, and thereby demonstrate clearly the need for deeper and more serious attention to Martian beauty than has generally been attempted (either by artists or, I submit, by those of us interested in the ethics of terraforming). In allowing our view of Mars to be distorted by a geocentric aesthetic attitude, we have failed to recognize and promulgate those distinctive qualities of Mars that five decades of in situ photography have already revealed, qualities at which I have gestured in this chapter. In *Red Mars*, after Sax Russell's extended aesthetic defence of terraforming Mars, Ann Clayborne's response is straightforwardly compelling: "You've never even seen Mars" (Robinson 2001, 175).

15.5 Concluding Remarks

If aesthetic objections to terraforming are to be dismissed on the grounds that Martian beauty is valuable only in proportion to its resemblance to the more familiar natural beauty of Earth, or that the beauty of a terraformed Mars would for other reasons be as valuable or more valuable than the beauty of Mars today, this judgment had better be the result of a deep and serious engagement with existing Martian aesthetic qualities. I have argued that this kind of engagement has not yet been generally practised, and that consequently the distinctive beauty of Martian landscapes has been widely overlooked in favour of "terraformed images". Some readers may be inclined to think that no aesthetic response to Martian landscapes is more valid than any other. I disagree; my own view aligns with that of Ronald Hepburn, the philosopher who led the revival of environmental aesthetics in the twentieth century. Hepburn claimed that the aesthetic appreciation of nature, like that of art, occurs at higher or lower levels partly reflecting the knowledge and attitude of the witness. Suppose we behold an ancient rock-face, he wrote. "Compare [...] the realizing of the pressures, thrustings and great age of the rock before us, with merely chuckling over the likeness of its markings to a funny face" (Hepburn 1966, 305). Aesthetic appreciation of the former kind is "less superficial or contrived", "truer to nature" and "for that reason more worth having".

To the extent that a serious, true-to-nature aesthetic engagement with Mars is possible today, I think it must involve, firstly, close and open-minded attention to the images of Martian landscapes currently available for study and, secondly, the imaginative contemplation of these images in light of the known facts about Mars. The character and quality of aesthetic experience, especially of natural scenery, is often strongly determined by contextual knowledge. To give another example due to Hepburn, the aesthetic response to the Andromeda galaxy viewed through a telescope is very different from the response to an abstract painting that happens to contain the same patterns of light and dark: "My awareness that the first shapes are of enormous and remote masses of matter in motion imparts a strangeness and solemnity that are not generated by the pattern alone" (ibid., 301). In a similar way, our response to Martian landscapes is modified by our awareness of their remoteness, their antiquity, and especially their uniqueness, the very distinctness from Earth that terraforming would obliterate.

It seems quite possible, I must admit, that as Mars becomes less remote and less strange, its landscapes will lose their lustre and come to seem ordinary or even ugly. Of course, they may yet reveal unimagined sensory delights. A consensus may emerge about the beauty of Mars, just as it has, broadly speaking, about the beauty of Earth. Or it may not; we cannot yet say. But if distinctively Martian forms of beauty are really there to be discovered (and, again, it seems to me that they are), it is likely that future exploration will bring them fully into public consciousness long before terraforming becomes a serious prospect. Aesthetic responses to Mars will shift unpredictably in the decades and centuries to come, as the planet is transformed from a remote and quasi-mythical space into a familiar, humanly intelligible place—even a home.

The global political movement to protect Earth's places of great natural beauty, which began in the nineteenth century and led to the creation of national parks all over the world, was spearheaded by those fortunate men whose aesthetic sensibilities had been attuned by prolonged immersion in the wild (John Muir, Henry David Thoreau, and others). Such immersion in the wilds of Mars has not yet been possible for us, and I do not think we know enough about Mars to conclude with any finality that its aesthetic treasures are more valuable than the possible rewards of terraforming. But by studying Mars I have come to recognize beauty of an unearthly kind in its pale colours and pure, ancient landforms. On the face of it, the destruction of this beauty would seem to be a terrible loss, and this alone provides a clear, if defeasible objection to terraforming.[3]

References

Cockell, C. S., & Horneck, G. (2004). A planetary park system for Mars. *Space Policy, 20*, 291–295.

Hepburn, R. W. (1966). Contemporary aesthetics and the neglect of natural beauty. In B. Williams & A. Montefiore (Eds.), *British analytical philosophy* (pp. 285–310). London: Routledge and Kegan Paul.

Herschel, W. (1784). On the remarkable appearances at the polar regions on the planet Mars, the inclination of its axis, the position of its poles, and its spheroidal figure; with a few hints relating to its real diameter and atmosphere. *Philosophical Transactions of the Royal Society of London, 74*, 233–273.

McKay, C. (1990). Does Mars have rights? An approach to the environmental ethics of planetary engineering. In D. MacNiven (Ed.), *Moral expertise: Studies in practical and professional ethics* (pp. 184–197). London: Routledge.

Protocol on Environmental Protection to the Antarctic Treaty. (1991). October 4, 1991. 30 ILM 1461.

Robinson, K. S. (2001). *Red Mars*. London: HarperCollins.

[3]I thank Evan Rodriguez, Diana Marosi, and the editors for insightful criticism of earlier versions of this chapter.

Rolston, H, I. I. I. (1986). The preservation of natural value in the solar system. In E. C. Hargrove (Ed.), *Beyond spaceship earth: Environmental ethics and the solar system* (pp. 140–182). San Francisco: Sierra Club Books.

Schiaparelli, G. (1888). Ueber die beobachteten Erscheinungen auf der Oberfläche des Planeten Mars. In M. W. Meyer (Ed.), *Himmel und Erde, 1 Jahrgang* (pp. 1–17). Berlin: Hermann Paetel.

Schwartz, J. S. J. (2013). On the moral permissibility of terraforming. *Ethics and the Environment, 18*, 1–31.

Sparrow, R. (1999). The ethics of terraforming. *Environmental Ethics, 21*, 227–245.

Sparrow, R. (2015). Terraforming, vandalism and virtue ethics. In J. Galliott (Ed.), *Commerical space exploration* (pp. 161–180). London: Ashgate.

Viking Lander Imaging Team. (1978). *The Martian landscape*. Publication SP-425. Washington, D.C.: NASA History Office. http://history.nasa.gov/SP-425/cover.htm

Wells, H. G. (1898). *The War of the Worlds*. London: William Heinemann.

Part V
Ethical and Legal Issues in Solar System Exploration

Chapter 16
'The Way to Eden': Environmental Legal and Ethical Values in Interplanetary Space Flight

Christopher Newman

16.1 Introduction

There is growing academic attention being directed towards the technical and engineering issues that face human spaceflight in respect of a journey to Mars. Often overlooked in such discussions, however, are the legal and ethical underpinnings that are needed to regulate human behaviour on such an extraordinary mission. This chapter will examine key issues in respect of law and ethics that will face humans on a long duration spaceflight. Given that there is an emerging consensus on the damage caused by existing space activities, it is contended that such a regulatory framework should embed a commitment to environmental protection within the fabric of rules governing human behaviour in outer space. The discussion will examine the extent to which legal regulation should be used to embed environmentalism as a core mission value. It will be argued that the harm caused by environmental damage should be viewed as serious enough to warrant criminal sanction, but will ultimately question whether this will be the most effective way to embed an ecological ethic into the behaviour of crews who travel to other planets.

When considering the nature of crimes that may be committed in space, simply discussing laws to which these space travelers should be bound to adhere would unhelpfully limit the scope of this discussion. Ethical considerations include breaches of social conventions that extend far more widely than criminal sanctions. Yet, even when attempting to define what activity should attract a criminal sanction on Earth, criminal theorists encounter difficulties. Criminality is, after all, merely legal classification of certain acts that are prohibited and attract a punishment. As Ormerod and Laird identify "...the nature of the act in question, its morality or immorality and the consequence do not change overnight; but its legal nature does" (2015, 4). Deciding what acts should attract this legal liability is no easy task but

C. Newman (✉)
Sunderland Law School, University of Sunderland, Sunderland, UK
e-mail: chris.newman@sunderland.ac.uk

J.S.J. Schwartz and T. Milligan (eds.), *The Ethics of Space Exploration*, Space and Society, DOI 10.1007/978-3-319-39827-3_16

criminal law theorists generally accept that there is no one "single concept or value that will capture the essential characteristic in virtue of which crimes are properly punished" (Duff 2004 quoted in Ormerod et al. 2014, 4). This chapter will not attempt to draft criminal laws for space, merely to highlight the areas for concern that could attract such criminality.

The chapter will begin by establishing the legal status of travellers on long duration, interplanetary travel. Whilst a trip to another planet is unprecedented, there are models for regulation of behaviour that can be utilised from the agreements that established the International Space Station, and these will be discussed. Although some work has been done in respect of the impact of long duration space activity upon human behaviour, this chapter suggests that the regulation of long distance, interplanetary space missions may need to differ from the arrangements currently in place to cover activity in Low Earth Orbit (LEO). This is for a number of reasons. First, there will be necessary autonomy that a crew engaged in long duration flight away from Earth will need, where real-time communications with Mission Control are not possible.[1] Second, given the diverse nature of longer duration or interplanetary missions, the nature of the space traveller recruited for such missions may be very different from the ones currently recruited for human spaceflight. Additionally, there may be physical and psychological issues that are amplified when travelling between the planets in the solar system that are either ameliorated or not felt with current human space activity.

These factors, taken together, mean that legal and ethical values currently at play cannot be automatically transposed onto these new types of missions. The chapter will examine the problems that could be encountered in the administration of justice away from Earth and the nature of criminality in space. The discussion will show that, as envoys of mankind, the most serious harm that an individual astronaut could cause during an interplanetary voyage is damage to the space environment. Whilst immeasurably small, the "tragedy of the commons" that has been seen in respect of space debris in earth orbit must not be allowed to be replicated in outer space. The work will conclude with an evaluation of how best to ensure that ecological values are embedded as a lodestar for astronaut behaviour.

16.2 Unpicking the Astronaut Conundrum: Participants in Space Flight

There has been much conjecture amongst the space law community as to the legal status of space travellers. Much of the discussion has revolved around the distinction between astronauts and space tourists (see, for example, Freeland 2010). It is argued, however, that such demarcation is not sophisticated enough for the challenges of

---

[1]For the time delay between Earth and Mars and the resultant difficulties see: http://blogs.esa.int/mex/2012/08/05/time-delay-between-mars-and-earth/.

interplanetary travel. As Harrison (2001) suggests, the first crew traveling to Mars will be a multi-skilled, protean mixture of engineers and scientists with rudimentary medical training. The legal framework surrounding the space travelers is very much rooted in the Cold War architecture of the Outer Space Treaty 1967.[2] As Gabrynowicz (2004) and Blount (2012) both identify, the primary purpose behind the drafting of the OST was to ensure security for the two cold war power blocs. It is unsurprising, therefore, that a central element of that far reaching international treaty, and the subsequent progenitor treaty, the Rescue Agreement of 1968, was focused upon the role of participants in the highest profile theatre of Cold War conflict.

The fragility of the technology in the early years of human space activity led to concerns that that there was a real danger, upon re-entry, that the space objects could land in a different part of the world from which they were launched. Von Der Dunk (2008) highlighted that, this being the case, it was feared that there was a real risk of the craft and the crew being recovered by a hostile nation. Von Der Dunk classified the Rescue Agreement as having three main elements: "(1) astronauts in distress on earth should be assisted as much as possible, (2) astronauts in outer space should be equally assisted as much as possible and (3) states are generally obligated to provide information that will aid such assistance" (Von der Dunk 2008, 416). This may seem a sensible precaution, especially given the historical context in which the treaties were drafted. Sundhal (2009) recognizes, however, that the OST and subsequently the Rescue Agreement are both limited in respect of the protection that they afford and inconsistent in the application of that protection. The OST affords protection to 'astronauts' as 'envoys of mankind'. The Rescue Agreement, however, eschews these terms speaking of 'personnel' and omitting reference entirely to the concept of envoys of mankind.[3]

The distinction between 'astronauts' and 'personnel' is, it must be noted, a largely arbitrary one. It is clear that participants in long duration, interplanetary missions would be classed as astronauts for the purposes of the OST and Rescue Agreements. Such a finding is significant for three key reasons. First, in the terrestrial phase of the mission, they would enjoy the protection afforded by the Rescue Agreement. This protection also would extend in the deep space phase of the mission, although (given present technology) this would be of little practical utility. Second, 'astronauts' are also legally recognised as being 'envoys of mankind'. A crucial feature of any interplanetary mission would be the notion that those

[2]The 1967 Treaty on Principles Governing the Activities of States in the Exploration and Use of Outer Space, including the Moon and Other Celestial Bodies (Herein referred to as the Outer Space Treaty or OST) was adopted by the General Assembly of the UN on 19 December 1966 by virtue of Resolution 2222 (XXI). It was opened for signature on 27 January 1967 and entered into force on 10 October 1967. It can be found here: http://www.unoosa.org/oosa/en/ourwork/spacelaw/treaties/introouterspacetreaty.html.

[3]Sundhal (2009, 178) indicates that as the Rescue Agreement was clearly intended to build upon the relevant provisions of the OST, the omission of the phrases 'astronaut' and 'envoys of mankind' intend to broaden the scope of the duty to rescue and assist to encompass such ventures as space tourism.

making the journey would represent all of humanity. This is significant when considering the scope and nature of behaviour that may attract the stigma of criminality. Finally, the term 'astronaut' identifies a uniformity of legal status for travellers in deep space. There will be no ambiguous hierarchy that might have existed between those who categorized themselves colonists or scientists or engineers. This uniformity provides a conceptual foundation upon which the notion of equality before the law can be extended to space travellers.

16.3 Criminality in Space: To the ISS and Beyond

It has been identified that long duration and interplanetary spaceflight poses unique issues in respect of identifying and codifying criminality. For the foreseeable future, the crew of any such flight will be relatively small. Each individual member of the crew will serve a specific purpose and have a specific expertise. "There will be a minimal social order and the burden of determining what infractions do or don't constitute a crime may fall on a group of individuals living and working in close proximity" (Redfield 2004, 80). Those charged with maintaining discipline in space will have a balancing act to make. On the one hand, there is the harmony and amity that needs to exist between such a group living and working at close quarters. There is, however, a need to ensure that the deliberate and malicious actions of an individual do not endanger the well being of the group and success of the mission.

The experience of managing astronauts during long duration visits to the International Space Station (ISS) at least provides a template for managing humanity in space. Governance arrangements for this complex international venture operate on a number of different levels. The ISS itself was established by means of an Intergovernmental Agreement (IGA) signed by all co-operating States in Washington in 1998. There are an additional four memoranda of understanding (MOU) that are signed between NASA and each Partner State and then there are various implementation agreements. The IGA is, therefore:

> …a structure of rules which can be considered as a framing law, since it considers more or less specific details for general questions, while it refers more specific questions such as crew management, to regulating documents on the matter specifically established for the Space Station. (Sgrosso 2004, 65)

Contained within the IGA, at Art. 22, are specific provisions relating to criminal activity on board the ISS. Rather than follow the OST regime of liability,[4] the IGA

[4]Article VII of the OST 1967 provides that "Each State Party to the Treaty that launches or procures the launching of an object into outer space, including the moon and other celestial bodies, and each State Party from whose territory or facility an object is launched, is internationally liable for damage to another State Party to the Treaty or to its natural or juridical persons by such object or its component parts on the Earth, in air or in outer space, including the moon and other celestial bodies."

chooses to import the criminal law of the state of the nationality of individual astronauts and bind them accordingly. Art. 22(1) states inter alia that the signatories to the IGA may exercise criminal jurisdiction over personnel in or on, any flight element who are their respective nationals. This means that if a Russian were to commit a crime against a Japanese astronaut in a module that was registered to the United States, responsibility for criminal jurisdiction would be Russian. Having established this Art. 22(2) then provides for the level of transgression that will attract prosecution. Specifically, it will be in a case involving misconduct on orbit that (a) affects the life or safety of a national of another Partner State or (b) occurs in or on or causes damage to the flight element of another party.

The basis for the administration of justice and the instigation of prosecution therefore, is the nationality of the perpetrator of the alleged infraction. As has been recognized, this means that each of the Partner States "have to consider measures not only to actually exercise their criminal jurisdiction but also to categorize the crimes which will be targeted for prosecution" (Farand 2004, 76). Art. 22(2) also has a provision whereby the Partner State that has suffered due to a transgression by the national of another Partner State can exercise jurisdiction if the national's partner state has failed to examine *the possibility of laying charges* (emphasis added) and there has been a mandatory consultation. Clearly, this is a relatively low hurdle for a Partner State to overcome, making this possibility somewhat remote. Additionally, as Hermida (2006) points out, there is no reference made in the IGA to issues of concurrent jurisdiction. This would normally be dealt with via treaty arrangement rather than customary international law and it is unclear what the position would be where two or more courts from different states can claim jurisdiction over a specific case. It should be noted, it is extremely unlikely that this would ever occur and the existence of this provision serves more as a procedural safeguard to engender confidence in the process.

The general principles of prosecution outlined in Art 22 of the IGA, are augmented with a separate Code of Conduct for the International Space Station Crew (CCoC) drafted via the Multilateral Coordination Board (which, of itself, was established by the IGA and MOUs which created the legal framework for the operation of the ISS). Part I (B) defines the scope and content of the CCoC as being applicable to all ISS crewmembers from the time they are assigned to the crew by the competent astronaut management body (Farand 2004). Art 11.6 of the MOU provides that the CCoC will inter alia establish a clear chain of command on-orbit, clear relationships between ground and on-orbit management, set forth disciplinary regulations, and provide the ISS Commander with appropriate authority and responsibility to enforce safety, security and crew rescue procedures. The CCoC introduces a disciplinary policy that binds the individual crewmember to a number of regulations that cover each stage of the mission. The CCoC also refers to stand alone policies in respect of ISS training, Earth to Orbit transfer vehicle and ISS flight rules. According to Farand,

> the interest of this disciplinary policy lies in the implicit recognition by cooperating agencies that their astronauts' behaviour may be subject to a process that is administered not only on the basis of their own personnel policy but also of the rules developed by the ISS partnership. (2001, 65).

Both the IGA provisions relating to criminal prosecutions and the CCoC for the ISS are germane to this discussion because they provide an indication of the way in which international partnerships currently address the issue of criminality and crew conduct on long duration missions. The 'sub-contracting' of prosecutorial discretion to the state of the accused bypasses the need for a detailed criminal code in space and provides an obvious route for managing any disciplinary issues. It is also clear that normative behaviour is inculcated from the embryonic stages of the mission: recruitment, selection and training. By selecting people who are harmonized with the collective goals of the mission and ensuring the crew is compatible from the outset, the potential for criminality and broader conflict is greatly reduced. In relation to their status under the OST, there is little doubt; the mandatory training requirements, rigour of selection and the altitude satisfy Lyall and Larsen's conditions for the status of astronauts within the terms of Art V of the 1967 Treaty (Lyall and Larson 2009). Accordingly, the ISS crew would be afforded protection under the Rescue Agreement of 1968. Whilst this is somewhat self evident, it does serve to illustrate that training and selection is a mission critical element of long duration spaceflight. Any legal and ethical code for interplanetary travel needs to ensure it starts to bite at the primary stages of mission development as opposed to commencing in space.

The state-based approach adopted in respect of the ISS and criminality, however, suffers from a fundamental limitation in respect of transposing this model to interplanetary travel; the space activity in question, whilst being long duration, is still occurring within Low Earth Orbit (LEO). The transit time from the ISS to the surface of the Earth is around three and a half hours (Wright 2015). The dominant view from the observation windows on the ISS is that of Planet Earth. Herein lies the crucial difference between the administration of justice on long duration space flight and an interplanetary mission. The next element of this discussion will, therefore, focus some of the issues that those engaged upon interplanetary flight will encounter and how these crucial differences between long-duration stays in LEO and travel between planets may require a different approach to the regulation of such space travellers.

16.4 The Urban Spaceman: At Home in the Hostile Environment

There are a number of problems facing humans who seek to voyage through the solar system to another planet. Many of these issues, focusing on the issue of the landing upon and colonization of, an alien planet (see Listner and Newman 2015) are beyond the ambit of this discussion. There is, however, a significant knowledge deficit in respect of the human ability to cope with the unique difficulties of a long duration spaceflight away from the relative closeness of LEO (for further information on the paucity of research on this see Nicolas et al. 2013). If, as Duque (2004) identifies, the job of an astronaut is exceptional in a number of respects, then the job of an interplanetary astronaut is without parallel. As Manzey (2004) states, a mission to another planet, such as Mars, will be unlike any other human undertaking. When considering the broad ethical underpinning of a regulatory system for interplanetary travel, the stresses on human beings involved in spaceflight must, therefore, be taken into account.

Harrison (2001) identifies four key ways in which spaceflight environments differ from those on Earth. One of these elements, as already discussed, is the depravation of the amenities of every day life. Comparisons with explorers from history and journeys of exploration on planet Earth only work up to a point given that the space environment is fundamentally inimical to human life. In addition:

> There are tremendous discontinuities between then and now… Unlike early seafarers, who traded hardship at home for hardship at sea, spacefarers must make major sacrifices in their accustomed quality of life. (Harrison 2001, 31)

The nature of the technology and the unforgiving harshness of space mean that at any point in the mission a failure of a human or a technical failure can result in death to all crewmembers. This is not unique to long duration, interplanetary travel; the aforementioned CCoC is predicated on ensuring that both the Ground Control and Station Commander have sufficient authority to ensure the safety of the station. Even so, the dynamics of interplanetary travel are very different in this respect. The ISS has the two Soyuz-TMZ capsules attached upon which the two teams of three arrived and this can be separated from the ISS in around 3 min (Moskowitz 2011). Provisions and spare parts are shipped regularly up to the ISS from Earth via the Progress modules (NASA 2015). This is a crucial difference between the long-duration stay on the ISS and a long duration, interplanetary flight, where there would be no such safeguards protecting the crew.

Two of Harrison's four environmental considerations in relation to spaceflight are kindred in nature and particularly germane to discussions on interplanetary travel; those of isolation and confinement. Research has confirmed that psychological and psychosocial factors can significantly affect human behaviour during spaceflights and other extreme environments (Bishop 2004). Due to the distance from the Earth, interplanetary travellers will not have the luxury of real time communication with friends and family. The only direct communication these astronauts will have will be with other members of the crew, all of whom will be suffering the same isolation issues. Chambers (2013) identifies that even experienced and highly trained astronauts will experience some side effects of isolation. This may become apparent particularly when missing cultural celebrations or even something as mundane as family birthdays. Similarly, confinement will be a phenomenon that will see travellers enclosed in the same, small habitat in the company of a very small group of fellow astronauts.

Manzey identifies that these significant psychological challenges relate to the restricted range of environmental cues, the repetitive nature of the work and the complex psychosocial situation "which is characterized by a lack of privacy, enforced social contracts with other crew members and separation from the usual social network of family and friends" (2004, 782). Whilst the manifestation of the effects of isolation, confinement and mundane yet vital tasks will vary depending upon the individual concerned, any attempt to criminalize or regulate behaviour will need to recognize that such factors will impact upon all crewmembers. No amount of regulation will be able to prevent behaviours driven by these psychological factors and to penalize crewmembers will result in unnecessary stigmatization and perhaps even drive crewmembers away from seeking early intervention.

In the midst of such considerations, Chambers (2013) highlights one of the crucial elements; the potential lack of access to mental health support services. This is surely an area where positive regulation in respect of recognizing the challenges posed by mental health difficulties can help in the recruitment, selection and training of astronauts. It is also an area where the CCoC and IGA for the ISS remain resolutely silent. Manzey (2004) also identifies that existing spaceflight studies alone, including the experiences of ISS crewmembers, will not be sufficient to accumulate the knowledge necessary to deal with psychological problems that will arise on an interplanetary spaceflight. Ground based studies and simulations may provide some insight but, if one of the key values underpinning space activity is safety of the human participants, there will need to be more consideration of the protection of the mental health of the crew. It will not be sufficient to merely hope this happens. Regulation and imposing duties on the mission planners, ground control and those on the voyage (including a duty to include someone within the crew trained to a competent level as a counselor) may well be a significant, if unseen, contribution to the well being of an interplanetary, crewed mission.

16.5 Master and Commander? Administration of Justice in Space

This discussion has established the mechanics of how regulation of crew behaviour is currently maintained on long duration space flight. As has been seen, codes of conduct for space travelers are the principal tool by which the behaviour of astronauts is currently regulated. Working alongside such codes, however, there is also the (admittedly unlikely) threat of legal sanctions, which are in place for serious breaches of behaviour on board the ISS. The remote nature of interplanetary travel means the deferral to terrestrial jurisdiction is neither desirable nor practical. Any mechanism for determining normative conduct amongst space travelers will ultimately need to see a self-contained approach to the administration of justice, although how practical this will be for the foreseeable future is open to question.

The defining feature of a criminal law is that it is a prohibition accompanied by a sanction. The effective punishment of interplanetary travelers is an issue that is distinctly problematic. The uniquely remote nature of interplanetary space travel, coupled with the rigorous selection of the astronaut grouping, means that none of the usual models of criminological theory are effective at fully exploring or predicting the commission of crimes in space:

> …until criminology comes up with a thorough understanding of the causes of crime in outer space, the criminal justice system will lack the necessary theoretical tools to design a criminal justice approach to effectively deal with these conflicts. (Hermida 2006, before n. 135)

Particularly problematic is the mechanics of investigation and punishment and how they will be administered on an interplanetary space mission. It is the "prospect of punishment which differentiates criminal proceedings from other state-sponsored proceedings." (Wilson 2014, 51). This, however, brings us to the first definitional obstacle when considering the administration of justice in outer space; who should be considered 'the state' or sovereign authority and who should administer any resulting punishment. Classical constitutional theory holds that there should be a separation of powers,[5] with the legislative body, the executive authority and the judicial function being kept distinct and separate to prevent abuses of power (Bradley and Ewing 2014).

Such a separation of the investigative and judicial roles within the confines of an interplanetary space mission poses clear practical problems. Whilst popular culture (particularly within the Star Trek franchise) tends to give unlimited authority to the mission commander, this is problematic for two reasons. The first is cultural; the omnipotent commander approach tends to indicate a military structure with attendant codes of discipline. The experiences of the ISS would tend to suggest a deliberate attempt to emphasize the civilian, peaceful nature of space exploration. Indeed Article 1 of the IGA echoes Article I of the OST stating that the ISS shall be a civilian, international space station for peaceful purposes. Second, and from the perspective of administrating justice, concentrating the investigative function and judicial authority in the same position would clearly be contrary to the separation of powers principle that characterizes the majority of countries' constitutions.

As has been shown in respect of the ISS, the nationality of the astronaut provides the legal framework for the investigation and prosecution of criminal offences. This would seem to be an unsatisfactory long-term solution, although given the distances that humans can currently travel, it may be that this approach is likely to remain in place for the foreseeable future. Irrespective of which 'sovereign authority' is granted responsibility to investigate and adjudicate on criminal infractions by

[5]The theory of the separation of powers is particularly significant in respect of criminal justice as the executive will be comprised of the police, who are responsible for the investigation of crimes and they will work with prosecutors to build the case against a suspect. The need for the judiciary to be independent from the executive is a fundamental requirement of the rule of law. For further details of the operation from a UK perspective see Bradley and Ewing (2014).

crewmembers, there is a second element of punishment to consider. Here, the key issue is that:

> …as well as general questions of jurisdiction there are also questions of how to treat a suspect before trial, and what, if any, punishment to administer while in space. (Redfield 2004, 80)

Punishment is, perhaps, the most problematic area of contemplation for the administration of justice on interplanetary missions and one where there are few obvious solutions. Given current (and foreseeable) engineering capability, living space is likely to be at a premium on any interplanetary spacecraft. Redfield (2004) identifies that the nature of such a craft means that individual space travelers will already be confined in a general sense. Restrictions of space mean that it does not seem likely that provision for bespoke, isolated detention facilities will be made given that living space will be at a premium on board interplanetary craft. The principle tool of punishment (incarceration away from the main social group) would seem to be inappropriate and unworkable. In addition, such isolation may be impractical with the offender having duties vital to the success of the mission and key to the survival of the other crewmembers. Other punitive sanctions, such as financial penalties, may be appropriate but of limited utility on a mission that may take many years to complete.

The administration of justice, both in terms of investigation and punishment, poses a number of significant questions for long duration space travel away from Earth. Despite the previously identified differences between life in LEO onboard the ISS and conditions faced by interplanetary travelers, it is likely that the twin approach adopted by the ISS, (that of an overarching code, coupled with nationality-based jurisdiction) will be imported to any interplanetary mission. In the event of serious criminality a course of action will be determined on a case-by-case basis with reference to mission control on Earth. Whatever solution is adopted, it is suggested that the issue of criminality is considered proactively as part of the planning stage of an interplanetary mission. Otherwise mission control and those onboard the ship will be forced to reactively consider how to deal with an event that could adversely affect the mission.

16.6 Whose Crime Is It Anyway?

Whilst the issues of investigation and punishment pose difficult questions for future interplanetary travelers, the rigorous and intensive selection process for astronauts can provide a robust, in-built defence against criminality. Despite the zeitgeist-capturing ‘Mars One’ project utilizing an open selection, it is axiomatic to say that astronaut selection will remain highly competitive and selective (Norberg and Steimle 2013 at p. 255). Those with a criminal record are likely to be ineligible and the initial vetting of astronauts at the stage of application works in concert with an on-going programme of evaluation designed to ensure maximum compatibility with others in the astronaut group. Harrison (2001), drawing on the psychological studies in Antarctica of Edholm and Gunderson (1973), identifies three qualities which will undoubtedly reduce a propensity for criminal activity, specifically;

> space farers who are capable of sustained high performance, who are emotionally stable and who can get along with one another. (Harrison 2001, ch. 6 at n. 13).

All societies need a basic level of stability in order to function (Smith and Natalier 2005) and long duration space travelers will be no exception. In the closed community of an interplanetary space mission, this need for stability is underpinned by the task-orientated nature of space flight. An explicit hierarchy reinforces this. The commander of the mission is responsible for the safety of the crew and the success of the mission. Each crewmember will have responsibility for an aspect of the overall success of the mission. In addition, even the most mundane task [identified by Harrison (2001 at Ch. 9) as repairing life support systems, vacuuming filters and preparing food] assumes a mission-critical level of significance. Accompanying this will be a recruitment and selection process designed to ensure maximum compatibility within the group. As Hermida states:

> ...the recruitment and training programs are so demanding and place such an enormous emphasis on physical and mental conditions by keeping those that show "criminogenic" characteristics or proclivity to commit crimes out of the recruiting process. (Hermida 2006, after n. 95)

The selection process for interplanetary spaceflight is, inter alia, predicated on maintaining a closed, harmonious group where anti-social behaviour and criminality are minimized (Harrison 2001 and Maschke et al. 2011). This does not, however, eliminate the prospect of the commission of a crime. As Wilson (2014) recognises, the use of criminal law is designed to be coercive, laying down the barriers of behaviour that will be permitted and introducing punishments for transgression. This leads to the obvious question in respect of long duration interplanetary space travel as to what behaviour should be sanctioned as criminal.

Simester et al. (2013) have identified two broad conditions that should be met if behaviour is to be categorized as criminal and sanctioned accordingly. First, the activity concerned must be 'sufficiently serious' to warrant intervention, leading to significant levels of harm being suffered by others. If this duty is a positive one, to safeguard others, the second criteria acts to constrain; the criminal law must be shown to be the most preferable form of legal control. Crucially, they add one important rider that; "the practicalities must be considered of drawing up an offence in terms that are effective, enforceable and meet the rule of law" (Simester et al. 2013, 646).

The idea that crimes should be introduced only when they are needed to prevent harm runs throughout criminal theory. This is founded on the notion of autonomy, as articulated by Mill in *On Liberty*, which holds that authorities may only punish to prevent harm to others. In normal (terrestrial) society, the corollary of this is that, beyond the constraints of the harm principles, individuals are free to behave in any way they see fit.[6] In the confines of an interplanetary space mission, it is clear that such a concept will in practice be subservient to the mission planning and the roles and responsibilities of crew members within that plan. Given the hostile nature of outer space and the requirements of the mission, there will be restrictions imposed on the behaviour of the crew. Not all of these will be criminal, instead being enforced by ‘softer’ codes of conduct and more informal, social processes within a code of conduct. It is recognised that there is only a remote prospect of criminal activity being a serious problem on interplanetary missions. Well-trained and carefully selected astronauts should share a ‘common professional discipline’ (Redfield 2004, 80) that means criminality is likely to be restricted to the realms of the improbable. If the approach adopted by the ISS is extended to other forms of space exploration, and the criminality of the astronaut is contingent upon her or his nationality, there will be a clear body of jurisprudence with significant areas of commonality regarding laws preventing offences against the person, murder, sexual offences and property transgressions.[7]

Notions of crime and punishment in respect of interplanetary missions can only partially be dealt with via existing legal and regulatory regimes. The governance of astronauts during long duration visits to the ISS is bespoke to the agreement that established the Station. It is also predicated on the relative proximity of the ISS to Earth. The unique human habitat of a spacecraft used for interplanetary space travel means inevitably that there would need to be fresh consideration of any crew code of conduct, even if the basic principles of liability remain. Establishing the need for a revivified regulatory regime, therefore, requires consideration of the broader space environment. It is contended that the type of harm that human activity could cause to the delicate, pristine environment of outer space is potentially more far reaching than any other consideration, including harm to other crewmembers. This discussion will, therefore now examine the need to embed environmental protection within the fabric of rules governing human behaviour in outer space.

[6]Criminal theory has long discussed the notion that conduct should only be criminalized if it results in harm to another. This “Harm Principle” was first articulated in John Stuart Mill, *On Liberty* (1859). This principle was developed to include occasions where serious offence to others arose from the conduct in Feinberg, *The Moral Limits of Criminal Law* (1984–1987); vol. 1, *Harm to Others* (1984); vol. 2, *Offense to Others* (1985); vol. 3, *Harm to* Self (1986); vol. 4, *Harmless Wrongdoing* (1988). It is not the purpose of this discussion to analyze these theories. For contemporary exposition of the competing issues see, Simester, Spencer, Sullivan and Virgo, *Simester and Sullivan's Criminal Law* Chapter Sixteen, *The Moral Limits of Criminalization*, pp. 643–669; Wilson (2014), Ashworth (2009).

[7]Art 22 of the IGA requires that each signatory to the IGA have to make sure that there is sufficient provision within their respective legal systems to categorize the crimes that are likely to be targeted for prosecution (Farand 2004 at p. 76).

16.7 Preserving the Pristine Environment: Ecologically Driven Regulation

In considering notions of 'harm' and the foundation of criminal liability, the damage caused by human space activity is an area to which one is unavoidably drawn. There has been much written about the environmental impact of space activity. It is an area that has seen, perhaps, the most significant shift in normative values over the course of human exploration of space within the twenty-first century. Williamson (2006) and Viikari (2008) both explore the fragility of the extra-terrestrial environment and conclude that human activity has contributed to the degradation of the space environment threatening its sustained use. Contemporary discussions on the environmental impact of humans on outer space have tended to focus on the question of dealing with space debris (Hobe 2012) and also potential damage to celestial bodies due to speculative and commercial mining (Newman 2015a). Nonetheless, the actual process of interplanetary travel will not be without an environmental impact. That it will be immeasurably small, does not mean that it should be overlooked. As Welly (2010) points out, early advances in space technology meant that human-made debris was not considered as being problematic; yet now such debris is regarded as a serious threat to on-going human space activity.[8] The situation of interplanetary travel cannot be discounted as posing similar risk.

One of the key difficulties in this area is the lack of explicit, hard-law attention to environmental issues in outer space. This is a critical area of oversight, however; that utilitarian and pragmatic approaches to space activity have sidelined any attempts to provide meaningful new treaties (Newman 2015b). This has led to a manifestation of the 'tragedy of the commons' where individual countries have received benefits from individual missions whilst damaging the 'global commons' (Welly 2010, 279). A lack of bespoke legally binding environmental provisions has created a largely unfettered space environment. It is voluntary and non-binding codes of conduct tend to predominate in environmental space regulation. The UN Debris Mitigation guidelines cover the area of space debris (UN Office of Outer Space Affairs 2010).

The threat to other celestial bodies from human activity is even more fundamental than that posed by space debris. Accordingly, there are guidelines in respect of planetary protection (Robinson 2006). Once an alien environment has been contaminated, be it by a robotic probe or by human settlers, the environment has been corrupted irrevocably. This will have two key impacts; first it will limit scientific study by providing a false positive reading of life (Butler 2006). Second, there may be profound implications for any life that does exist should humanity contaminate a biosphere. This is outside the area of contemplation for this inquiry, but it is nonetheless an area that will need consideration as interplanetary travel draws closer. When considering notions of harm, therefore, it is not only planetary

[8]For consideration of the legal implications of Space Debris and especially issues regarding liability for damaged caused by debris, see Listner (2012).

colonization and space debris that has an environmental toll. It is germane to this discussion to consider the act of interplanetary travel and how this may have an ecological impact. Robinson (2006) states that there is an increased awareness of the importance of mitigating the environmental damage caused by human space exploration. It is at this point that a dilemmatic choice emerges; is the harm caused serious enough to justify a criminal sanction, with the attendant difficulties relating to the administration of justice in outer space that such an approach begets?

There is clear justification for sanctioning ecological harm to the space environment within criminal theory. Feinberg (1984, 11) states that harm refers to a wrongful setting back to some protected interest, leaving that interest in a worse condition than it otherwise would have been. Transposing this to astronauts on an interplanetary space journey, the degradation of space would be seen as leaving the protected interest of the space environment as being in a worse condition. It is argued that prohibiting such harm would help in the balancing of pragmatic realities against environmental imperatives. This echoes similar ethical debates in other areas of environmental regulation, specifically surrounding the issue of climate change. The Hardinian 'tragedy of the commons' outlined above means that to ignore the threat of environmental damage caused by space travel is ultimately to risk the safety of space travellers and the sustainability of future space activity. Criminal behaviour might include a prohibition of ejecting anything from the craft, imposing legally binding duties on crewmembers with responsibilities for craft subsystems that might leak substances into space and a specific regulation dealing with waste management on the journey.

The advantage of an enumerated list is that clear parameters can be set, detailing the nature of prohibited conduct. The principle of *nullum crimen, nulla poena sine lege*[9] embedded in most legal systems holds that any criminal law must be drafted so that an individual knows when they are committing an offence. The obverse, however, is also true; enumerated lists of discrete offences means that anything committed outside the scope of that list will be acceptable. Imposing a general duty on crewmembers would, therefore, 'future proof' the underpinning environmental ethic of a code of conduct, making it more pervasive than lists of offences which may well be made obsolete due to advances in space technology. It may also be that criminalizing discrete elements of environmental damage is not only unnecessary but also, ultimately, counterproductive.

The preferred option to ensure such harm may be mitigated is by embedding environmental protection within a crew code of conduct on interplanetary missions. As stated above, 'softer' codes of conduct rely on more informal, social processes that emphasize common values and shared goals. Those forming the code of conduct might wish to impose specific conditional criteria, detailing exactly the type of damage that should be prevented. There may, however, be a more general statement imposing a duty on all crewmembers to minimize environmental damage caused by interplanetary travel. The CCoC underpinned by a general environmental duty would

[9]Literally translated as meaning 'no crime, no punishment without law'.

remove the adversarial nature of apportioning blame for an infraction and instead embed a duty to minimize environmental damage on all aspects of the mission. The history of space governance shows that early patterns of regulation tend to be sustained as normative behaviour. Given the earlier observations about the discipline of professional astronauts, it is suggested that the creation of a duty to safeguard the wider space environment could have enduring benefits for future space travel.

16.8 Conclusion: To Infinity and Beyond?

A human journey to another planet will soon become a reality. Such a journey poses unique challenges for the regulation of the behaviour of the crew. Astronauts will carry the legal burden of being envoys of mankind, a role as yet undefined but one that is sure to carry more weight in the context of interplanetary travel. The rules governing conduct on the ISS, both the use of nationality to assign criminal liability and the CCoC governing behaviour from first selection provides a template for regulation. Given the differences between activity in LEO and interplanetary travel, this is only going to be a short-term solution. As technology develops and human capacity to travel moves beyond Mars, new solutions will need to be found regarding the administration of justice. There will need to be an effective separation of the investigation of crimes in space from the judicial process. Punishment away from Earth will also need to be considered.

More fundamentally, the basis of what behaviours should be criminalized in outer space will need to be addressed. This inquiry has shown that the harm caused to the delicate space environment by human activity could necessitate criminal sanction. The 'tragedy of the commons' that has already beset LEO should not be replicated in respect of interplanetary space travel. It is suggested, however, that instead of longer-term criminalization, the creation of an overarching environmental value system, within a revivified CCoC would embed ecological protection within the planning, training and the day-to-day tasks of astronauts on board interplanetary spacecraft. Making environmental protection a core value of those who travel to other planets would provide a meaningful lodestar for them as envoys of humanity. Establishing environmental protection as the normative position for astronauts will reorient human space activity. Effective regulation on these early journeys of exploration to another planet may yet show humanity the way to Eden.

References

Ashworth, A. (2009). Chapter Two, Criminalization. In *Principles of criminal law* (6th ed., pp. 22–43). Oxford: OUP.

Bishop, S. (2004). Evaluating teams in extreme environments: From issues to answers. *Aviation, Space and Environmental Medicine, 75*(Supplement 1), C14–C21.

Blount. (2012). Renovating space: The future of international space law. *Denver Journal of International Law and Policy*, *40*, 515.

Bradley, A., & Ewing, K. (2014). *Constitutional and administrative law* (16th ed., pp. 88–95). London: Pearson.

Butler, J. (2006). Unearthly microbes and the laws designed to resist them. *Georgia Law Review*, *41*, 1355–1394.

Chambers, C. (2013). Mars one: The psychology of isolation, confinement and 24-hour Big Brother. *The Guardian*. [online]. Available at: http://www.theguardian.com/science/head-quarters/2013/sep/09/neuroscience-psychology. Accessed December 6, 2015.

Duque, P. (2004). Life in Space. In *ECSL, ESA, UNESCO Joint Symposium on the Legal and ethical framework for astronauts in space sojourns* (pp.79–81). [Online]. Paris: UNESCO. Available at: http://unesdoc.unesco.org/images/0013/001397/139752m.pdf. Accessed December 5 2015.

Edholm, O., & Gunderson, E. (1973). *Polar human biology*. London: Heinemann Medical.

Farand, A. (2001). The code of conduct for international space station crews. *ESA Bulletin*, *105*, 64–68.

Farand, A. (2004). Astronauts' behaviour on board the international space station: Regulatory framework. In *ECSL, ESA, UNESCO Joint Symposium on the Legal and ethical framework for astronauts in space sojourns* (pp. 70–78). [Online]. Paris: UNESCO. Available at: http://unesdoc.unesco.org/images/0013/001397/139752m.pdf. Accessed December 6, 2015].

Feinberg, J. (1984). The moral limits of the criminal law.

Freeland, S. (2010). Fly me to the Moon: How will international law cope with commercial space tourism? *Melbourne Journal of International Law*, *11*, 90–118.

Gabrynowicz, J. (2004). Space law: Its Cold War origins and challenges in the era of globalization. *Suffolk University Law Review*, *37*, 1041.

Harrison, A. (2001). *Spacefaring*. Berkeley: University of California Press.

Hermida, J. (2006). Crimes in space: A criminological and criminal justice approach to criminal acts in outer space. *ERCES Quarterly Review*, *2*(2).

Hobe, S. (2012). Environmental protection in outer space: Where we stand and what is needed to make progress with regard to the problem of space debris. *Indiana Journal of Law and Technology*, *8*, 1–10.

Listner, M. (2012). The space review: Addressing the challenges of space debris, part 2: Liability. [online]. *Thespacereview.com*. Available at: http://www.thespacereview.com/article/2204/1. Accessed January 10, 2016.

Listner, M., & Newman, C. (2015). Failure to launch: The technical, ethical, and legal case against Mars One. [online]. *Thespacereview.com*. Available at: http://www.thespacereview.com/article/2712/1. Accessed December 6, 2015].

Lyall, F., & Larsen, P. B. (2009). *Space law*. Farnham, England [u.a.]: Ashgate Pub.

Manzey, D. (2004). Human Missions to Mars: New psychological challenges and research issues. *Acta Astronautica*, *55*, 781–790.

Maschke, Oubaid, V., & Pecena, Y. (2011). How do Astronaut candidate profiles differ from airline pilot profiles? *Aviation Psychology and Applied Human Factors*, *1*(1), 38–44.

Moskowitz, C. (2011). *How would NASA and Russia evacuate the international space station*? [online]. *SPACE.com*. Available at: http://www.space.com/12772-nasa-space-station-crew-evacuation-plan.html. Accessed DecEMBER 6, 2015].

NASA. (2015). *Russian progress spacecraft*. [online]. Available at: http://www.nasa.gov/mission_pages/station/structure/elements/progress.html#.VmRO8YTndHw. Accessed DecEMBER 6, 2015.

Newman, C. (2015a). Seeking tranquillity: Embedding sustainability in lunar exploration policy. *Space Policy*, *33*, 29–37.

Newman, C. (2015b). The undiscovered country: Establishing an ethical paradigm for space activities in the 21st century. In A. Lawton, L. Huberts & Z. van der Wal (Eds.), *Ethics in public policy and management* (1st ed., PP. 299–324). Routledge.

Nicolas, M., Sandal, G., Weiss, K., & Yusupova, A. (2013). Mars-105 study: Time-courses and relationships between coping, defence mechanisms, emotions and depressions. *Journal of Environmental Psychology, 35*, 52–58.

Norberg, C., & Steimle, H. (2013). Chapter 7: Astronaut selection and training. *Human spaceflight and exploration*. Berlin: Springer.

Ormerod, D., Laird, K., Hogan, B., & Smith, J. (2014). *Smith and Hogan's criminal law* (13th ed). Oxford University Press.

Redfield. (2004). On the values Frontier: Ethics and criminal conduct in outer space. In *ECSL, ESA, UNESCO joint symposium on the legal and ethical framework for astronauts in space sojourns* (pp.79–81). [online]. Paris: UNESCO. Available at: http://unesdoc.unesco.org/images/0013/001397/139752m.pdf. Accessed December 5, 2015.

Robinson, G. (2006). Forward contamination of interstitial space and celestial bodies: Risk reduction, cultural objectives and the law. *ZLW, 55*, 380–399.

Sgrosso, G.C. (2004). Legal aspects of the astronaut in extravehicular activity and the "space tourist". In *ECSL, ESA, UNESCO Joint Symposium on the Legal and ethical framework for astronauts in space sojourns* (pp.79–81). [Online]. Paris: UNESCO. Available at: http://unesdoc.unesco.org/images/0013/001397/139752m.pdf. Accessed December 5, 2015.

Simester, A., Spencer, J., Sullivan, G., & Virgo, G. (2013). *Simester and Sullivan's criminal law* (5th ed.). Hart.

Smith, & Natalier, K. (2005). *Understanding criminal justice*. London: SAGE.

Sundhal, M. (2009). The duty to rescue space tourists and return private spacecraft. *Journal of Space Law, 35*, 163–200.

UN Office of Outer Space Affairs. (2010). *UN-space debris mitigation guidelines*. Vienna: UNOOSA http://www.oosa.unvienna.org/pdf/publications/st_space_49E.pdf

Viikari, L. (2008). *The environmental element in space law*. Leiden: Martinus Nijhoff Publishers.

von der Dunk, F. (2008). A sleeping beauty awakens: The 1968 rescue agreement after forty years. *Journal of Space Law, 34*, 411–434.

Welly, N. (2010). Enlightened state interest: A legal framework for protecting the "Common Interest of All Mankind" from Hardinian Tragedy. *Journal of Space Law, 36*, 273–313.

Wilson, W. (2014). Chapter Two, Decision to criminalize. In *Criminal law* (5th ed., pp. 34–51). Longman.

Williamson, M. (2006). *Space: The Fragile Frontier*. Reston, VA: American Institute of Aeronautics and Astronautics.

Wright, J. (2015). *Soyuz landing*. [online]. NASA. Available at: https://www.nasa.gov/mission_pages/station/structure/elements/soyuz/landing.html. Accessed December 6, 2015.

Chapter 17
The Risks of Nuclear Powered Space Probes

Paul R. Graves

17.1 Introduction

On January 28, 1978, Komos 954, a nuclear powered Soviet spy satellite, fell from orbit, scattering radioactive debris across a wide area of northern Canada. Given the reliability of launch vehicles and satellites, such uncontrolled, high velocity re-entries can only be completely avoided by getting out of the space business. Therefore if we are to continue using nuclear powered satellites and probes, we must consider the risks and moral implications of those risks. In this paper I will review some of the facts about nuclear powered space probes and consider the ethical issues raised by such probes. I will argue that on utilitarian grounds the risks from a small number of such missions is sufficiently small that they are morally defensible. I will argue on deontological grounds that the imposition of small risks on others is probably unavoidable and that again, the aggregate risk from a small number of missions falls below the level of significant moral concern.

17.2 Factual and Historical Overview

Since the early days of the space race, a small fraction of satellites and deep space probes have used nuclear decay to supply electrical power and to keep critical systems warm. Most of the fuel has been used in radioisotope thermoelectric generators, or RTGs. In an RTG, the fuel generates heat through nuclear decay. The concentration and mass of the fuel is below critical mass and therefore cannot melt down or explode (NASA 1999, p. 2). The RTG is a very simple device with no moving parts above the atomic level. The nuclear fuel generates heat through the

P.R. Graves (✉)
Department of Philosophy, Oakland University, Rochester, USA
e-mail: graves@oakland.edu

J.S.J. Schwartz and T. Milligan (eds.), *The Ethics of Space Exploration*, Space and Society, DOI 10.1007/978-3-319-39827-3_17

radioactive decay. The outsides of the RTGs usually have fins to radiate heat. The fins are colder than the fuel. This allows thermocouples to generate electricity from this temperature gradient. Although a variety of fuels have been used in RTGs, for spaceflight applications, the most common is plutonium 238,[1] in the form of plutonium dioxide, PuO_2. Plutonium 238 is not the isotope that is used in atomic bombs so there is little threat of RTGs being appropriated by parties seeking to make such weapons (NASA 1989, Sect. 2, p. 8).

RTGs are a particularly attractive type of power supply for deep space probes. They have a high energy density. That is they produce a great deal more electricity per kilogram in comparison with chemical batteries or fuel cells. Plutonium 238 produces approximately 0.4 W/g of fuel or roughly 4.0 W/cm^3 of fuel (JPL 1994, Appendix B, p. 3). This is important because it costs thousands of dollars per kilogram to put material into space. Beyond the cost, given the limitations of launch vehicles, more weight devoted to the power supply means less weight devoted to scientific instruments and therefore a decreased scientific return on a mission. RTGs work reliably for decades and in the space environment require no maintenance. The major alternative electrical power supply for long-term space missions is solar panels. But since sunlight falls as the square of the distance from the Sun, solar panels must increase in size in geometric proportion to the distance a spacecraft travels from the Sun. Since Saturn is roughly ten times as far away from the Sun as Earth is, a space probe orbiting Saturn would require solar panels roughly 100 times larger in area than the solar panels required to power the same spacecraft in Earth orbit. Solar panels also become less efficient in lower light levels so the 100 times larger solar panels is a conservative estimate (ibid., Sect. 4, p. 9; p. 11). Large arrays of solar panels can also block instruments and so require more maneuvering of the spacecraft to keep instruments pointed at scientific targets while the solar panels remain pointed at the Sun as much as possible. This in turn reduces the scientific return from the mission in comparison with a probe powered by RTGs (ibid., Sect. 4, pp. 18f). To date, no probe aimed beyond Jupiter has used solar panels. RTGs have been used in some Earth orbiting satellites, six Apollo Lunar Surface Experiment Packages, including one that returned to Earth with Apollo 13's lunar module (NASA 1989, Sect. 2, p. 14), the Viking I and Viking 2 Mars landers, The Mars Science Laboratory "Curiosity" rover, Pioneers 10 and 11, Voyagers 1 and 2, Galileo, Ulysses, Cassini (JPL 1994, Appendix C, p. 5) and New Horizons. These last 8 probes all went to Jupiter or beyond.

RTGs in spacecraft do present one risk that is not shared by other power supplies: The danger of releasing radioactive material in the event of an accident. NASA, the Soviet Union and now Roscosmos have gone to some lengths to minimize these risks. For example, after the satellites' useful missions were completed, both NASA and the Russian space agencies have placed nuclear power supplies in very high

[1]According to the Jet Propulsion Laboratory, "[r]adioisotopes other than Pu-238 have been used in RTGs for ground-based applications, although none have ever been used for U.S. Space missions" (JPL 1994, Sect. 4.1.1).

disposal orbits where they can be expected to remain until their nuclear fuel has decayed to safe levels. Indeed, it was the failure of such a system that led to the inadvertent re-entry of Kosmos 954 mentioned above. In addition to parking RTGs in stable orbits far from Earth, NASA has also hardened the fuel blocks in the RTGs to prevent or minimize the spread of radioactive material in the event of accidents. The plutonium oxide is encased in graphite. The graphite is then encased in an iridium skin which is expected to survive most re-entry scenarios. This is then encased in a carbon fiber shell which is again expected to survive the heat and forces of atmospheric entry (NASA 1989, Sect. 2, p. 14; Sect. 4, p. 9). Several of these fuel blocks are then combined together in the RTG housing.

A failed launch of the Nimbus 1 weather satellite in 1968 supports NASA's contention that the RTGs can be expected to survive the forces from a booster failure encountered early in flight. The rocket launch vehicle went out of control and had to be destroyed. In such a scenario, the RTG is subjected to the heat from the exploding rocket fuel only momentarily and the debris produced by the explosion is not traveling at a high enough velocity relative to the RTG to pose significant dangers of damaging the fuel blocks. Moreover the RTGs are sturdy enough to survive an impact in water. Nimbus 1's RTG was recovered intact from the floor of the Pacific Ocean and its fuel was subsequently used on a later satellite (ibid., Sect. 2, p. 14).

NASA asserts that the greatest dangers would seem to be from an explosion and fire immediately before or after launch, or an uncontrolled atmospheric reentry ending with the RTG fuel blocks striking rocks. If a fully fueled rocket were to explode and burn on the launch pad or in the first moments of flight, the RTGs could be exposed to intense heat for a protracted period of time. The heat from the fire could conceivably result in a failure of the RTG's fuel blocks and a release of the plutonium fuel. In such a scenario, any plutonium released would be expected to be confined almost entirely to the launch facility, e.g. Cape Canaveral (ibid., Sect. 4, p. 32). Cleanup activities at the launch site would reduce the risk of further plutonium exposure to very low levels. Even so, people working at the launch site, particularly fire fighters and cleanup workers, could be exposed to the plutonium fuel, resulting in some increase in cancer risk.

In the worst case scenario, a spacecraft flying past the Earth in a gravity assist maneuver would instead strike the Earth with a considerably higher re-entry velocity than from Earth orbit. In a gravity assist maneuver, a spacecraft passes close to a planet to use the planet's gravity to accelerate the spacecraft. This allows larger probes to be launched to distant planets with smaller booster rockets. To date Venus, Earth, Jupiter, Saturn and Uranus have all been used for gravity assist maneuvers. In particular, in both the Galileo and Casini missions, the Earth was used in gravity assist maneuvers involving RTG powered spacecraft. In each case the spacecraft came within a few tens of thousands of miles of the Earth. Such a maneuver poses a slight danger that the spacecraft might accidentally strike the Earth and at significantly higher velocities than in a fall from orbit or a failure to reach orbit. In such a scenario the RTG housings could rupture, scattering the fuel blocks. The fuel blocks would be exposed to high temperatures from atmospheric reentry, weakening or breaching some of the fuel blocks and then all of the fuel blocks could strike rocks on impact. Such an impact could rupture an RTG fuel block that had been structurally compromised by the higher, interplanetary reentry velocity. The resultant release of plutonium could be expected to cause a small increase in the global cancer rate. NASA estimated that the individual risk of cancer death from the launch of the Galileo mission to orbit Jupiter was on the order of one in one hundred million (ibid., p. v). It compared this to other possible causes of death, such as being struck by lightning, as one in few million. That is, lightning is roughly two orders of magnitude more deadly to a given individual than the risk of radiation exposure from the flight of the Galileo mission. In the final pre-flight environmental impact report, the tiny risk from flying the Galileo probe was weighed against the $800,000,000 that had already been spent on the project and the prospect of gaining scientific knowledge from the flight. More specifically, the environmental impact report lists the goals of the mission as

1. To further the understanding of the origin and evolution of the Solar System
2. To further the understanding of the origin and evolution of life
3. To further the understanding of Earth by comparative studies of the other planets. (ibid., Sect. 1, p. 1)

As a philosopher, such accidents interest me because the risks they impose on the population are very small but real. How should we evaluate small but real risks?

To begin with, if NASA's numbers are correct, the risk really is very small. Nor is there any reason to be particularly skeptical about NASA's numbers. I am not an engineer and so my opinion on the question is of little value, but my reading of NASA's environmental impact statements suggests that their estimate of the cumulative risks is quite conservative. That is, in stating that the individual risk to the general public from the Galileo mission was on the order of one cancer death in one hundred million, they appear to me to be, if anything, overestimating the risks. There are many more ways of missing the Earth than there are of hitting it, even if the probe is flying past Earth in a gravity assist maneuver.[2] Moreover, the implicit increase in cancer risk is, in the global context, probably undetectable. An extremely unlikely worst case accident might cause several thousand additional cancer deaths over the course of several decades, or a few hundred additional cancer deaths per year. Compare this with the World Health Organization's estimate of roughly 8,200,000 cancer deaths for 2012.[3] The additional cancer deaths from a failed space mission appear to be well within the margin of error in the estimates of global cancer deaths. That is, looked at globally, the increase in cancer deaths attributable to the worst case scenario would be undetectable in the global cancer statistics.

[2]In the interests of full disclosure, I should confess that I am a space enthusiast. I am happy to accept a one in one hundred million risk of a fatal cancer in order to get knowledge about the universe.

[3]http://www.who.int/mediacentre/factsheets/fs297/en/, retrieved August 3, 2015.

On the other hand, although the risk is small, it is not zero. If the expected additional cancer deaths from an RTG powered space probe is really on the order of one in one hundred million, then in a world with five billion people at the time of the Galileo launch or seven billion people now, the expected rate of fatalities is several dozen per flight. Is the information gleaned from (e.g.) the Galileo mission worth the expected cost of several dozen lives? So here at last we have the ethical puzzle: According to one way of thinking about the risk to human lives from a space probe, even the worst case accident, cannot be expected to measurably increase the cancer rate. The additional risk imposed on the world's citizens is truly negligible. But on the other hand, the average expected increase in cancer deaths per flight is a few dozen. Is it acceptable to conduct an experiment with an expectation of killing a few dozen people in order to acquire information about the outer planets of the Solar System?

17.2.1 Comparisons with Other Risks

NASA points out that the risk of cancer death from spacecraft RTGs is quite small. It is, for example, more than seven orders of magnitude smaller than the background risk of death from cancer. It is two orders of magnitude smaller than the risk of being killed by lightning. However there are morally significant differences between risks from spacecraft and risks from natural phenomena such as lightning. The risks from lightning differ from the spacecraft RTG risk in that lightning is a natural phenomenon. People do not cause lightning strikes. No one bears moral responsibility for a lightning strike. There are also simple countermeasures to protect oneself from a lightning strike, such as moving indoors when a storm approaches. In contrast, the risks imposed on people from spaceflight are caused by human choices and human actions. Unlike storm clouds, people bear moral responsibility for the consequences of their activities. So the people who design, build and launch spacecraft are morally responsible for the consequences of their actions. Moreover the countermeasures for protecting oneself from radiation exposure in the event of an accident are more involved than staying indoors until the storm passes.

Another complication is that the cancers that result from the spacecraft's nuclear materials cannot be distinguished from cancers from other causes. That is, for a particular cancer patient it will usually be impossible to determine whether that patient's cancer resulted from exposure to the spacecraft's nuclear fuel or from some other cause. This is a significant difference with some of the cases that are being used for comparison. When a person is struck by lightning, we know which individual has been struck. When a person is injured or killed in an automobile accident, we typically know that this particular person died as a result of the accident. In the case of a person who dies of cancer, we have no way of knowing whether the patient's cancer was caused by radioactive material from a spacecraft's RTG or by some other cause. We could make only a statistical estimate of the deaths caused by the spacecraft's RTGs. So the risks are so small even in the worst case scenario of a high velocity reentry accident, that they probably will not be detectable either in individual cases nor in the aggregate.

17.3 Utilitarian Considerations

To reiterate, NASA's estimated risk of fatality from the Galileo mission was estimated at 1 fatality expected in 100,000,000 (one hundred million) people. It can legitimately be asked whether expected deaths is really the right way to think about the risks. The expected deaths represent something like what we would expect the average number of deaths would be in a very long series of launches. But it is unlikely that we will have any large number of RTG powered spacecraft in flight any time soon. NASA estimates the probability of the worst case scenario accident happening at something in the neighborhood of one in one million (NASA 1999, p. 3). That is, they estimate that out of one million Earth gravity assist spaceflights, approximately one would accidentally hit Earth. To date, two such missions using RTGs have been flown. If we continue to fly one mission every decade, it would take 10,000,000 (ten million) years to fly one million missions, roughly one of which would be expected to accidentally strike the Earth in a gravity assist maneuver. If this is correct, then the risk of the worst case scenario from one mission or any small number of missions is vanishingly small.

Almost all of the risk seems to be concentrated in two phases of the flight: A not terribly unlikely launch accident which might cause dozens of cancers, and a very unlikely accidental impact during a gravity assist flyby maneuver that might cause thousands of cancers. The latter risk might be thought to be so rare and our experience so limited that we cannot really assess the risk with any great accuracy. Perhaps with probabilities this low and with a very small number of flights, a utilitarian should be willing to write the risk off completely.

An explosion and fire at launch is probably the most likely bad outcome. A few percent of our rocket launch attempts end in catastrophic failure. But even here there is a wide variety of scenarios. A rocket that explodes more than a few seconds into flight exposes the resulting fragments to high temperatures of combustion for a very short time as the fuel and rocket components rapidly disperse. The RTG from Nimbus 1 has been shown to survive such energies. To expose the RTG to prolonged intense heat from a fire requires a fully fueled rocket to fail on or very near to the launch pad. In such a case any leaked plutonium would be confined to a small area, limiting the number of people exposed and therefore limiting the cancer risk (NASA 1989, Sect. 4, p. 32). The overall risks imposed on the general population seem very small indeed.

The benefits of flying these missions are somewhat more abstract. Of course developing, constructing and flying these spacecraft employs many aerospace engineers, technicians and scientists. Were there to be no such missions to the outer planets, these engineers, technicians and scientists would have to find other work. It has also frequently been argued that a vigorous space program has economic benefits beyond the aerospace industry in the form of technological spinoffs. There is little doubt that this has been true of the space program to date. The scientific objectives of these missions are to learn more about the outer planets and their satellites. It has been said without exaggeration that these missions to the outer planets have produced far more information about them than everything we learned about them prior to the missions. While such information as we have so far acquired about these bodies has little immediate practical application, it is fair to say that basic research has a tendency to be of practical value in the long run.

Many would argue that the knowledge gained is of value for its own sake. Full disclosure: That is my opinion. The knowledge I have gained of Jupiter and Saturn from these spacecraft is well worth the tiny amount of my tax dollars that they have cost me to date and the tiny risks they have imposed on me. But I can certainly understand a less daring person who might deny that the knowledge gained is worth the risk imposed. So the utilitarian calculus comes down to weighing the current careers of engineers, technicians and scientists, plus the knowledge gained of the outer planets, plus technological spinoffs and economic benefits from the programs against the tiny risks imposed on the world's population.

Since my life is not infinitely valuable, it would seem to follow on utilitarian grounds that some quantity of gains in scientific knowledge and employment by scientists, engineers and technicians as well as spinoffs into the broader society and the potential, however vague, for future benefits would justify imposing some level of risk on the world's inhabitants. I am willing to accept these risks for myself in light of the proposed benefits of nuclear powered deep space probes, but other people might weigh the utilities differently.

17.4 Deontological Considerations

Deontologists maintain that the foregoing cost-benefit kind of analysis is the wrong approach to deciding ethical questions. Instead deontologists advocate attentiveness to moral duties and to due recognition of the special value of moral beings. Moral beings as such are entitled to special consideration in our choice of actions. They are not mere chits to be traded against other interests. Their value and autonomy is to be respected in all of our actions. So a deontologist may ask if it is permissible to expose moral beings to minute risks that flow from our actions, or if we have a duty to protect people from even very small risks that result from our actions?

17.4.1 Consenting to Risks

One widely recognized way of legitimately imposing risk on another is through consent. If a person understands and voluntarily undergoes a risk then this is acceptable. So on the advice of my physician I might accept the risks of surgery in order to repair an injury. I might even consent to an experimental treatment regimen, perhaps with greater risks or less certain risks than a more established treatment. The requirement of consent recognizes my inherent value as an autonomous moral being. I am entitled to decide whether to accept the risks or forego the risks.

In some cases there may be some fairly explicit consent to the risks that come with building and flying nuclear powered spacecraft. People who work in the nuclear industry preparing the fuel for the RTGs are presumably cognizant of the risks and assume them in exchange for wages. Technicians preparing the spacecraft for launch are presumably cognizant of the risks and accept those risks in exchange for wages. Firefighters and cleanup crews that would have the responsibility of removing contaminated materials in the event of a launch-pad accident presumably understand the risks and accept them as a condition of their employment. All of these people are presumably giving fairly explicit consent to the risks they are subjected to in doing their jobs.

However, for the vast majority of the people it is manifest that no such explicit consent exists in the case of nuclear powered spacecraft. We have not asked people to sign consent documents attesting to our understanding the risks and indemnifying space programs against injuries, disease or death resulting from accidents with nuclear powered spacecraft. In the case of nuclear powered spacecraft, few if any people have given explicit consent to undergo the risks imposed on them by the spaceflight.

It might be argued that consent has been given by proxy. As citizens of a representative government and through the electoral process, we consent to be governed, to abide by the laws and to accept at least some degree of shared responsibility for the actions of our government. Moreover, at least in the United States we have political rights including speech, petition and public demonstration. If we disapprove of our government's actions, we can protest, sign petitions, vote people out of office and so on. We also require environmental impact reviews to assess these risks as accurately as possible. These environmental impact reviews are intended to provide guidance to agency leaders and higher governmental authorities on the risks imposed by our nuclear powered spacecraft. Through the political process we charge our leaders with making reasonable assessments of the risks and acting in the overall interests of the citizens. The long-term popularity of NASA and the United States space program might be thought to imply consent to the risks imposed upon us by this program.

Suppose we stipulate that the United States citizens have collectively, through the political process, consented to the risks from nuclear powered spacecraft. Is that enough for the imposition of risk to pass deontological muster? Unfortunately the answer must be no. Because the worst case scenario risks are more or less equally shared among all people, even if participation in the political process by United States citizens implies consent to these risks, only a small fraction of the global population can be thought to have consented through the political process. United States citizens, after all, make up less than 1/20th of the world's population. The very great majority of the global population lacks even indirect political input into spaceflight decisions and therefore cannot be thought to have consented to the risk through the political process. Hence, if the imposition of risk is to have a deontological defense, it cannot be in terms of the political process.

17.4.2 The Inevitability of Risk

In laying out his deontological theory Kant maintained that we may not treat other rational beings *merely* as means to our ends. So I may not treat another person simply as a device for enriching myself nor exclusively as a means to my own pleasure without regard to the interests, autonomy and well-being of the person so used. However none of us are self-sufficient. We depend on others for many things such as food, water, clothing, shelter, education, medical care, toys and so on. So, (e.g.) my personal physician is a means for maintaining and restoring my health. For her I am a means of paying off her student loans and making a comfortable living for herself. If either of us treats the other *merely* as a means without regard to our special status as moral beings, this is morally wrong according to Kant. Ethical interactions, especially with those who are means to our ends, must be conducted with due respect to each other as moral beings. For me that means, among other things, paying my bills and treating my physician respectfully, perhaps following her recommendations for treatment at least inasmuch as these seem reasonable to me. For her it means among other things, treating me conscientiously with my best interests in mind and maintaining the customary confidentiality. Because it is impossible for us to live our lives without treating other people as means to our ends, treating people as means must be morally permissible. But it is only morally permissible when we do not treat people *merely* as means, but also as ends in themselves. A society of people engaged solely in mutual exploitation is not a moral society even if the mutual exploitation is consensual and mutually beneficial. A genuinely moral society is one where, in the process of using each other as means to our legitimate needs, we also treat each other with the respect and dignity to which all moral agents are entitled.

Perhaps something similar could be said with regard to the imposition of risk. Just as it is impossible for us to live in a complex (post-?) industrial economy such as ours without treating other people as means to our many ends, it is probably impossible for us to live in such a society without imposing risks on other people. If so a deontologist must recognize that it is permissible to expose other people to some small risks in the course of living together in a society. For example, in our ordinary social interactions we routinely expose each other to infectious diseases. We cough, we sneeze, thereby spreading infectious viruses and bacteria. We leave our bacteria behind on doorknobs, dishes, desks, machines and so on. These infectious agents impose risks on other people. In driving my car to work, I impose small risks on people driving other cars as well as people walking along the street and smaller risks on people as their distance from the street increases and they are protected by embankments, trees, bushes and walls. When I ride my bicycle to work

I impose similar risks on others. These risks are smaller because of the lower speed and lower mass of myself and my bicycle, but the probability of collisions may also be greater in that locally cyclists frequently share “safety paths” with pedestrians. Walking to work probably minimizes the risks I impose on others, but perhaps exposes me to greater risks than driving my car. The point here is that just as it is impossible for us to live our lives without in some measure treating people as means, it is also impossible for us to live our lives without imposing small risks on other people.

If the imposition of risk without consent is to have a deontological justification, it must be on the grounds that the risk is so small that imposing it does not compromise people’s well-being or autonomy in a significant or meaningful way. One way to make that case would be to argue, as NASA does in its environmental impact assessment, that the risks from the spaceflight are much smaller than risks that people routinely accept in living their lives. At one point NASA proposes a threshold risk as that of an airplane crash killing a person on the ground. If we can accept the risks imposed on us by airplanes flying overhead, we should accept the much smaller risks imposed upon us by spacecraft flying overhead. If I do not wrong someone by an action that imposes real but undetectably small risks on that person, perhaps that action is acceptable. If we choose to live in places where automobile accidents, earthquakes, tsunamis or tornadoes impose significantly greater risks of death than spacecraft, perhaps is it morally acceptable to fly the spacecraft.

17.5 Policy Recommendations

The preceding discussion suggests the following conclusions: I have argued on utilitarian grounds that in my view the economic and scientific benefits from flying nuclear powered interplanetary probes are sufficient to justify the very slight risks. I acknowledge that people of good will may disagree on this value judgement. I have also argued on deontological grounds that imposition of small risks on others must be morally acceptable because it is impossible for us to live our lives without imposing small risks on others. Still, in light of the fact that the risks from nuclear powered spacecraft are real but very small, and in light of the fact that few people can be thought to have given voluntary informed consent, it would seem that nuclear powered spaceflight should be minimized. At present supplies of Pu 238 and iridium are limited enough that very few spacecraft will be flown with RTGs such as those on the Galileo and Cassini probes in the foreseeable future.[4] It is my opinion that this tiny risk, iterated over a very small number of missions, imposes a deontologically acceptable increase in risk to the general population. That is, the

---

[4]NASA asserts that 1.6 % of US iridium reserves we dedicated to Galileo’s RTGs (NASA 1989, Sect. 4, p. 40).

incremental risk from these missions appears to be tiny in comparison with risks that we accept routinely as the inevitable risks of living and working together. Again I concede that people of good will may question this judgement. However, many such missions would have to be flown to make the risk to the general population significant. Most of the risk to those unaffiliated with the space program can be eliminated by banishing nuclear powered spacecraft from low Earth orbit. In view of the risks from fallback accidents such as Kosmos 954, nuclear power supplies probably should not be used in low Earth orbits where the spacecraft and their RTGs are likely to fall back to Earth eventually. On the other hand, once the spacecraft have left Earth orbit for other planets, the incremental risk each flight poses is microscopic. Moreover, there is at present no practical alternative to using RTGs to power our spacecraft flying in the outer solar system. Perhaps in a science fiction future where space is crowded with many nuclear powered spacecraft, a more stringent policy would have to be adopted. However, if, as I believe, such research is sufficiently important, then the minuscule risk it adds to our lives is justified.

References

JPL. (1994). *Cassini program environmental impact statement supporting study* (Vol. 2): Alternate mission and power study. JPL Publication D-11777, Cassini Document 699-070-2.

NASA. (1989). *Final environmental impact statement for the galileo mission*. http://ntrs.nasa.gov/archive/nasa/casi.ntrs.nasa.gov/19900002786.pdf

NASA. (1999). *NASA fact sheet: Spacecraft power for cassini*. http://saturn.jpl.nasa.gov/multimedia/products/pdfs/power.pdf

Chapter 18
Shaking the Foundations of the Law: Some Legal Issues Posed by a Detection of Extra-Terrestrial Life

Frans G. von der Dunk

18.1 Law and Ethics—Law as a Social Construct

In order to properly address the legal issues posed by a proper detection of extra-terrestrial life (not just a mere serious possibility, as with the recent discovery of actual water on Mars), because of its extraordinary character it is necessary to briefly revisit the foundations of 'the law' as a social construct, and explore its relationship to 'ethics' as another social construct.

The type of 'law' being discussed here is, of course, man-made, and made to deal with human activities, including human reactions to (other) events. Human-made law has for example been defined as "the principles and regulations established in a community by some authority and applicable to its people, whether in the form of legislation or of custom and policies recognized and enforced by judicial decision".[1] Further to such a general definition, human-made law is, essentially, meant to achieve two, sometimes cooperating, sometimes contradicting aims.

As most people would readily acknowledge, law is first supposed to establish some semblance of justice. What 'justice' is, is at least partially subjective and culturally determined, and moreover shifting over time. In order to properly adapt thereto, law also has its own inherent system for change—procedures for creating new treaties overriding old ones at the international level, procedures for creating new legislation overriding old legislation on the national level by parliamentary (or other) procedures.

The original version of the book was revised: The copyright in the source line of Chapter 18 was missed to be updated. The erratum to this chapter is available at 10.1007/978-3-319-39827-3_19

[1]As per http://dictionary.reference.com/browse/law.

F.G. von der Dunk (✉)
College of Law, University of Nebraska-Lincoln, Lincoln, USA
e-mail: fvonderdunk2@unl.edu

J.S.J. Schwartz and T. Milligan (eds.), *The Ethics of Space Exploration*, Space and Society, DOI 10.1007/978-3-319-39827-3_18

As a consequence, law presents a formalized and somewhat inflexible social construct as compared to, for example, social pressure, morals or, indeed, ethics (which has been defined for instance as "a system of moral principles"[2]). Law therefore usually tends to lag behind ethics; the more flexible and fluid character of the latter guarantees that the more formal process for changing the former takes more time, sometimes considerably so. Only if the paradigm changes in a certain area of morals or ethics are so fundamental in nature (and usually also within such a short timeframe) that the formalized procedures for changing the law are seen as part of the problem of the substantive law which is supposed to change radically instead of part of the solution, allowing for due adaptation of substantive law, will such procedures be ignored. This is essentially what we call 'revolution'.[3]

This is also why at a certain point in time the ethics with respect to a certain subject matter may not be completely commensurate with the law on such subject matter: the latter may not yet have caught up sufficiently—or there might be certain ethical principles which are too broad and too vague to give shape to meaningful legal requirements and obligations. As a consequence, whether a certain action or activity is 'ethical' or not only plays an indirect role in determination of its lawfulness: unless the law has fully and explicitly incorporated the ethical principle(s) at issue, the latter would serve not as a legal rule in itself but mainly as a guiding principle for helping interpret certain legal rules or principles. Oftentimes, thus, courts are encouraged to use such ethical principles as 'equity' or '*ex aequo et bono*' in applying the law. Still, the result may well be that at a given moment in time the law, or its implementation, within a given community does not always (completely) reflect the ethics of that community.

This is however where the second major role of the law comes in: it is *also* supposed to establish some efficiency in human interaction, to allow some measure of predictability of human action, thereby easing human interaction. This predictability is the root cause also of the inherent inflexibility noted; it usually precludes law from being changed on a whim, at the instigation of a single event or string of events without much thorough reflection and consideration. As a matter of fact, most law is developed more for this purpose than for achieving 'justice'—there is for example no inherent justice in driving left or right. This is also relevant for outer space, where there is not even a 'left' or 'right'.

A final key feature of the law as a consequence of its being man-made is that is essentially follows man: only when and where humans become active, might law become necessary—and do humans go through the trouble of developing it. In the international community, such an elaborate national legal order is fundamentally

[2]As per http://dictionary.reference.com/browse/ethics?s=t.

[3]Thus, 'revolution' has been defined as "an overthrow or repudiation and the thorough replacement of an established government or political system by the people governed", alternatively "a radical and pervasive change in society and the social structure, especially one made suddenly and often accompanied by violence"; as per http://dictionary.reference.com/browse/revolution?s=t.

based on state sovereignty over national territory: states are the legal 'constructs' of groups of humans through which public international law is developed and applied.[4]

18.2 Spanning the Globe: The Domain of International Law

Equally based on state sovereignty, on an international level law is fundamentally and foremost developed essentially *between* states, in the absence of any global legislative authority. Even the United Nations, often seen as the closest thing in the international community to a 'world government', can only in very exceptional circumstances impose its legal will upon unwilling states—namely if such a state has committed acts of aggression against other sovereign states or committed other very serious international crimes, and even then only if and to the extent that at least all five major powers holding the right of veto in the UN Security Council, agree.[5]

The most visible element of inter-state created international law consists of international treaties,[6] where states agree on draft texts ready for adoption, and then —on an individual basis—decide to adhere to or not, sometimes with individual conditions attached which further complicate the resulting legal situation. They can thus essentially determine which international obligations they are willing to accept in return for other states accepting the same obligations on a reciprocal basis.

In the present context, it could at least in theory be imagined that states would conclude treaties amongst themselves about joint approaches to the possibilities of finding extra-terrestrial life. However, states would presumably only go through such 'trouble' if an actual and urgent matter arises in this context—read: a major discovery of actual extra-terrestrial life, possibly intelligent, calling for immediate action—or in case of joint outer space projects searching for such life—in view of the presumably immense costs and technological capabilities this would require.

The second element, usually coming into play where treaties do not apply, or where their application is insufficiently clear, uniform and/or generally acceptable, is customary international law.[7] Customary international law in the last resort is

---

[4]See already Art. 2(1), UN Charter (Charter of the United Nations, San Francisco, done 26 June 1945, entered into force 24 October 1945; USTS 993; 24 UST 2225; 59 Stat. 1031; 145 UKTS 805; UKTS 1946 No. 67; Cmd. 6666 & 6711; CTS 1945 No. 7; ATS 1945 No. 1), which posits territorial sovereignty of states as perhaps the most fundamental tenet of the international legal order—even to this day.

[5]See Arts. 39, 41 & 42, UN Charter. The five states referred to are the United States, the Russian Federation, the United Kingdom, France and China.

[6]Cf. Art. 38(1)(a), ICJ Statute (Statute of the International Court of Justice, San Francisco, done 26 June 1945, entered into force 24 October 1945; 156 UNTS 77; USTS 993; 59 Stat. 1031; UKTS 1946 No. 67; ATS 1945 No. 1).

[7]Cf. Art. 38(1)(b), ICJ Statute.

about state behaviour and attendant state conviction that such behaviour is of a legally binding nature—as effectively interpreted by authoritative legal experts; if these generally speaking are not in agreement, most likely a case for the existence of relevant customary international obligations can not be made.

In the present context, it would be difficult to pinpoint any halfway realistic scenario whereby states, through their behaviour in a consistent manner *vis-à-vis* (the possibility of existence of) extra-terrestrial life and attendant convictions, establish a relevant rule of customary international law in terms of what types of actions they might be expected or even required to undertake, although it can of course not be completely excluded. One main element in customary international law is precisely the *customary* part; meaning usually a number of similar situations need to be analysed before anyone could come to the conclusion that a rule of customary international law has arisen in that context. So far, of course, we have encountered a bit of evidence only of the *likelihood* of particular extra-terrestrial life, and state action would most likely not anytime soon rise to the level of a consistent '(state) practice' from which legal convictions could be derived.

A final point concerns the concepts of 'subjects' and 'objects' of the law. The former term refers to those who are formally entitled to rights under a legal system *and* to defend themselves against interference with those rights. In terms for example of contract law and national law this concerns *inter alia* natural persons, with the exception of children and other persons considered mentally incapable. Like animals, they have certain rights, but are unable to defend themselves under the law. The latter term, of 'objects of the law', refers precisely to those latter categories.

In international law, however, traditionally only states qualified as subjects of the law. Increasingly it may also apply to intergovernmental organizations, private entities and individuals, but that is still exceptional and limited to certain domains only. For example, individual humans have been given certain independent standing in the field of human rights law, being allowed themselves to protect their interests in appropriate international courts and tribunals. In all other respects, however, intergovernmental organizations, private entities and individuals still remain mere objects of the law—the 'animals' and 'children' of international law.

18.3 Going Extra-Terrestrial: The Special Role of Space Law

Within the above context of general public international law, space law refers to a specific sub-set of legal rules applicable to outer space and human activities therein. Consequently it developed basically following the launch of Sputnik in 1957, as a relatively novel area of law.[8] The 1967 Outer Space Treaty[9] functions as the overarching document in this context, providing the general legal framework for all human activities in outer space, for the benefit of all mankind and for peaceful purposes.[10]

Further major principles include the responsibility and liability of states also for private activities in outer space, meaning that if, for example, in the course of the search for extra-terrestrial life, private associations or organizations violate applicable rules of international law or cause damage to other humans and their activities, whether on earth or in outer space, their states are to be held accountable at the international level.[11] Consequently, states increasingly develop national (space) laws including systems of licensing and authorization in order to take care of such accountability.[12]

Also with regard to intergovernmental organizations, while they are referenced in the treaty as major players in the space arena, ultimately the member states bear international responsibility and liability for any space activities conducted within their frameworks.[13] Consequently, states here would do well to—and usually indeed did—provide for an internal system to ensure that the organization would not undertake any activities which could thus entail their own responsibilities or liabilities.

States are furthermore to respect the freedom of exploration and use of outer space, *inter alia* targeting scientific exploration,[14] and respect other legitimate space activities by consulting with others if their own activities may result in harmful interference with those of others as well as accept requests for consultation by such others.[15]

Finally, Article III of the Outer Space Treaty ensures that wherever the treaty itself or other special rules of space law are not sufficiently clear or even absent, general public international law—notably the UN Charter, which is expressly

---

[8]Thus, the UN Committee on the Peaceful Uses of Outer Space (COPUOS) was established in 1958/1959 to discuss legal issues resulting from outer space activities and come forward with proposals for developing relevant law at the international level.

[9]Treaty on Principles Governing the Activities of States in the Exploration and Use of Outer Space, including the Moon and Other Celestial Bodies (Outer Space Treaty), London/Moscow/Washington, done 27 January 1967, entered into force 10 October 1967; 610 UNTS 205; TIAS 6347; 18 UST 2410; UKTS 1968 No. 10; Cmnd. 3198; ATS 1967 No. 24; 6 ILM 386 (1967).

[10]See in particular Arts. I-IV, Outer Space Treaty.

[11]Cf. Arts. VI, VII, Outer Space Treaty.

[12]Major examples thereof concern the United States, the Russian Federation, Ukraine, the United Kingdom and France, but also smaller states such as South Korea, Sweden, Belgium, the Netherlands and Austria.

[13]Cf. Arts. VI, XIII, Outer Space Treaty.

[14]See Art. I, Outer Space Treaty.

[15]See Art. IX, Outer Space Treaty.

mentioned in view of its key role in the context of international peace and security —can be used to delineate the relevant legal situation, rules, rights and obligations.

Further to the Outer Space Treaty itself, a few other treaties have been developed to elaborate on specific aspects of the former. Notably this concerns the 1968 Rescue Agreement,[16] the 1972 Liability Convention[17] and the 1975 Registration Convention.[18]

Effectively, the scope of the legal regime thus established (even if only in embryonic fashion) extends as far as human activities extend into outer space—the Outer Space Treaty itself in this respect as per most of its relevant Articles refers to "outer space, including the moon and other celestial bodies".[19] Informally, however, this clause is accepted by most experts as being limited to our solar system—likely and/or mainly because so far only very few man-made artefacts have left that solar system.

Another set of international rules, generally relevant for space activities yet usually distinguished as a separate regime, is that of telecommunications law, a sub-set of legal rules applicable to the international (effects of) use of radio-frequencies and, in the case of satellites, attendant orbital slots or orbits. A key role here is played by the International Telecommunication Union (ITU), which started to address satellite communications also shortly following the launch of Sputnik in 1957.

The ITU, on the basis of the ITU Constitution,[20] the ITU Convention[21] and the Radio Regulations[22] as overarching documents, most importantly takes care of the coordination of all international usage of radio frequencies.[23] This includes

[16]Agreement on the Rescue of Astronauts, the Return of Astronauts and the Return of Objects Launched into Outer Space (Rescue Agreement), London/Moscow/Washington, done 22 April 1968, entered into force 3 December 1968; 672 UNTS 119; TIAS 6599; 19 UST 7570; UKTS 1969 No. 56; Cmnd. 3786; ATS 1986 No. 8; 7 ILM 151 (1968).

[17]Convention on International Liability for Damage Caused by Space Objects (Liability Convention), London/Moscow/Washington, done 29 March 1972, entered into force 1 September 1972; 961 UNTS 187; TIAS 7762; 24 UST 2389; UKTS 1974 No. 16; Cmnd. 5068; ATS 1975 No. 5; 10 ILM 965 (1971).

[18]Convention on Registration of Objects Launched into Outer Space (Registration Convention), New York, done 14 January 1975, entered into force 15 September 1976; 1023 UNTS 15; TIAS 8480; 28 UST 695; UKTS 1978 No. 70; Cmnd. 6256; ATS 1986 No. 5; 14 ILM 43 (1975).

[19]Cf. e.g. Arts. I, II, III, Outer Space Treaty.

[20]Constitution of the International Telecommunication Union (ITU Constitution), Geneva, done 22 December 1992, entered into force 1 July 1994; 1825 UNTS 1; UKTS 1996 No. 24; Cm. 2539; ATS 1994 No. 28; Final Acts of the Additional Plenipotentiary Conference, Geneva, 1992 (1993), at 1.

[21]Convention of the International Telecommunication Union (ITU Convention), Geneva, done 22 December 1992, entered into force 1 July 1994; 1825 UNTS 1; UKTS 1996 No. 24; Cm. 2539; ATS 1994 No. 28; Final Acts of the Additional Plenipotentiary Conference, Geneva, 1992 (1993), at 71.

[22]Cf. Arts. 4(3), 6, ITU Constitution.

[23]See Arts. 1(2), 44, ITU Convention; Art. 7, ITU Convention.

listening to signals from and transmitting messages to extra-terrestrial life. Such coordination aims at the avoidance of (international) interference through a complicated system of allocation of frequency bands to categories of services and allotment of frequencies to certain systems. The activity of 'listening' to radio waves coming to earth from deep space—also known as 'radio astronomy'—has become recognized as one of those services, so as to enjoy interference-free 'use' of certain frequency bands duly allocated to it.[24]

18.4 Law and (the Search for) Extra-Terrestrial Life

Moving finally to the issue of 'law' *vis-à-vis* extra-terrestrial life, the first thing to note is that 'law' as such, including space law, does not address any extra-terrestrials. It is intended, as indicated, for actions of and/or interaction between *humans*, or at least their legal creations such as companies, associations and states.

This then also includes all kinds of human activities in the context of the search for extra-terrestrial life—not necessarily, however, as we shall see, extra-terrestrial life itself. Such activities, in theory, could be undertaken at three levels: private (that is, private associations and organizations, university and other research institutes—as least as long as not government-run—and individual persons), national-public (meaning governmental organizations, notably space agencies) and public international (meaning inter-state organizations, whether the European Space Agency (ESA) on behalf of its member states, or the United Nations on behalf of the world community). Legally-technically speaking, all three result in the responsibility and liability of one or more states, which would consequently presumably wish to have some level of control over such activities, even if they might condone them as such.

It is certainly too early to even consider establishing prohibitions at the international level to conduct such searches and efforts to establish contact by individual private organizations, certainly as long as condoned by the respective states internationally accountable for them, let alone search activities conducted by individual states. The sovereignty of states at the international level here translates into a fundamental freedom of *inter alia* action, expression and opinion for each such state,[25] unfettered by any international authority dictating and controlling such efforts or messages.

At the same time, of course, there would be an inherent logic in involving only a single channel for, or at least coordinate at a single level, substantial human activity in the search for extra-terrestrial life, in particular if intelligent—and even more in

[24]Cf. Art. 1.13, 1.58, Radio Regulations.

[25]While the sovereignty of states as perhaps the most fundamental structural rule of international law is reflected in many legal documents (including e.g. the aforementioned Art. 2(1)1, UN Charter), e.g. Art. 2(4), UN Charter, specifically protects political independence of any state against any forceful interference by another state.

particular if requiring either immediate action or otherwise large investments and a large technological know-how. Such efforts would then most likely not be guided by the more scientific international bodies such as COSPAR or the IAU, much as their scientific efforts would seem to lead the way and indispensable as such efforts might be. In view of the major political and social overtones of any game-changing discovery of extra-terrestrial life, such a role would seem to fit more logically a body representative not only of all relevant scientists of the world, but of all humans of the world, which in view of the current structure of the international community boils down to a body representative of all *states* of the world.

18.5 Mankind's Embassy? The Role of the United Nations

In other words: this is where the United Nations would most likely be drawn, or even invited into the game. Of course, also the United Nations has its inherent limitations and handicaps in such a context. It should, for example, be realized at the outset that it was originally established for reasons of political security, essentially to help prevent something as atrocious as the Second World War and the attendant crimes against humanity from ever recurring.[26] This might not necessarily make it a perfectly logical platform for scientific efforts to search for extra-terrestrial (intelligent) life, especially as long as the actual discovery of intelligent life would seem a remote (both in terms of time and in terms of chance) possibility.

On the other hand, already from the start 'political security' was interpreted broadly, as encompassing economic, social and legal security, and as time passed was relatively 'easily' further extended to medical, educational and even, nowadays, ecological realms of security. The United Nations consequently has served as a platform for the establishment of treaties dealing with environmental pollution and climate change, has established special agencies such as the FAO and UNESCO to deal with hunger and education respectively, and has also at a political level oftentimes allowed disputes to be solved or even pre-empted by peaceful means rather than by resort to armed force. From this perspective, the intended role for the United Nations indeed might well make sense: would not a realistic possibility of extra-terrestrial threats present the largest of all possible threats to (political/human) security?

The present structure of the organization is ruled by the UN Charter, which in the context of threats to international peace and security *inter alia* provides for the right of self-defense and a duty of international cooperation to counter any such threat.[27] Thus, states are individually or with the help of their allies entitled to defend themselves with force against an armed attack threatening their territorial independence and integrity—at least for as long as the international community

[26]See Preamble, UN Charter.

[27]See Art. 51, also Art. 1, UN Charter.

fails to take adequate measures to stop and roll back the attack and wipe out its consequences as far as possible.[28]

At the same time, in spite of its near-global membership[29] the limits of the role of the United Nations have also become clear: oftentimes, when states undertook actions in the international arena relying on the use of force which could not be brought under the relatively narrow and well-circumscribed terms of the UN Charter's clause on self-defense, they choose to justify such use of force by reference to a—by definition—much more vaguely and generally broader right of self-defense under customary international law. Inevitably, in many cases UN involvement has been politicized, often meaning the objectively most correct or desirable outcome would not be realized. Ultimately, it is the collective community of states which determines the extent to which the United Nations can, and would, actually take action in cases where, from an objective perspective, there could be little doubt that international peace and security are under threat.

In this context of international security, the Charter also established the main two organs of the United Nations and provides them with relevant competences.

The first organ is the General Assembly, representing states rather than mankind or individual peoples; the General Assembly is not an ordinary democratic institution.[30] Also this presents an important caveat to the desirability of UN involvement in the context of extra-terrestrial life issues, in particular if such an involvement would come to be exclusive. The key rule in this context is that of 'one state, one vote'; regardless of size of population or landmass, economic or military power, or of political, economic and social system, all states are at least formally speaking equal to each other. Furthermore, generally speaking the General Assembly can not take binding decisions; its powers are limited to debating and agreeing on Resolutions, which though politically important and equipped with the inherent ability to reflect or turn into customary international law, as such are not binding legal documents.[31]

This is in contrast with the second organ, the Security Council.[32] Here, some states turn out to be more equal than others, through the key role of the five permanent members (the United States, the Russian Federation, China, the United Kingdom and France) representing the reality of power politics at least at the end of the Second

[28]Art. 51, UN Charter, reads: "Nothing in the present Charter shall impair the inherent right of individual or collective self-defense if an armed attack occurs against a Member of the United Nations, until the Security Council has taken measures necessary to maintain international peace and security. Measures taken by Members in the exercise of this right of self-defense shall be immediately reported to the Security Council and shall not in any way affect the authority and responsibility of the Security Council under the present Charter to take at any time such action as it deems necessary in order to maintain or restore international peace and security."

[29]Currently, the United Nation counts 193 member states; see http://www.un.org/en/sections/about-un/overview/index.html, http://www.un.org/en/members/growth.shtml.

[30]See Arts. 9–22, UN Charter.

[31]Cf. Arts. 13–14, UN Charter.

[32]See Arts. 23–32, UN Charter.

World War.[33] As a consequence of this more 'realistic' dichotomy between the then-major powers and the other states, the Security Council may take binding decisions (in the limited area where it has to fulfill of its role, but this precisely includes issues of international peace and security) which may even allow for or in themselves include the use of force.[34] The composition of the Council, however, in particular the prerogatives of the five permanent members and their veto rights, is increasingly under fire as no longer properly representing the current power balance —such states as India, Japan, Germany and Brazil, and sometimes even the European Union as such, occasionally claim to have equal rights to such a permanent seat.

Specifically with respect to its role in outer space, the United Nations has established the Committee on the Peaceful Uses of Outer Space (COPUOS), as a committee of the General Assembly, in 1958. Currently, it comprises 77 states, more or less those most interested in outer space and space activities.[35] As a remnant from the Cold War, COPUOS works with consensus, down to determining the official agenda, which effectively excludes dealing with threats to the peace.

More generally, in the context of COPUOS space law has developed firstly by way of a handful of treaties in the late 60s and 70s, followed by UN Resolutions providing guidelines for certain types of space activities. COPUOS operates through two subcommittees, where—provided it would be accepted that the United Nations should take a leading role in this context—the Scientific and Technical Subcommittee may act as an embryonic international organization on space science in the context of any activities relative to extra-terrestrial intelligence (primarily the search therefore, so far), and the Legal Subcommittee might follow up with drafting guidelines for those activities. Ultimately, such political or legal measures would revert to the UN General Assembly for approval, after which they could be promoted to a UN Resolution—or even to a draft treaty, open for signature and ratification by individual states.

18.6 A Point of Reference for UN Action? The Case of Near-Earth Objects

An interesting reference point for such an approach is presented by the case of Near-Earth Objects (NEOs), which also present a kind of threat to earth from outer space. Here, the international initiative came from the Association of Space Explorers, a non-governmental private organization comprised by individuals which drafted a report that has since been fed into the discussions in COPUOS.[36]

[33]See in particular Arts. 23(1), 27(3), UN Charter.

[34]Cf. Arts. 39–46, in particular Art. 42, UN Charter.

[35]See http://www.unoosa.org/oosa/en/members/index.html, http://www.unoosa.org/oosa/en/COPUOS/copuos.html.

[36]See United Nations Report of the Committee on the Peaceful Uses of Outer Space, Fifty-third session (9–18 June 2010), A/65/20, §§ 136–142.

Amongst others, the proposals plan to use the General Assembly and Security Council in specific new roles fine-tuned to the case of NEO threats, whilst leaving the general structure intact.

Under these proposals, the General Assembly might essentially have a role in mandating, on behalf of all states and indirectly of mankind, the Security Council to coordinate (and undertake, as necessary) actions against NEO threats, in the process 'waiving' any liability for damage that might result as long as such actions take place within the mandate.[37] Such an approach thus recognizes the fact that only a handful of states are actually capable of undertaking substantive action once a NEO threat has been identified, and though the composition of the Security Council does not necessarily or comprehensively equate with those states, the five permanent members are certainly amongst the major space powers.

The precise relevance of the ongoing developments regarding NEO threat mitigation for the case of search for extra-terrestrial life, certainly if intelligent, may have to be analysed in greater detail, but they certainly show that, at least in principle, the United Nations could provide a roughly appropriate platform to use also in that context. It is essentially the only readily available platform of almost global scope, where member states have learned to some extent to cooperate together for the perceived greater common good of mankind.

However, one key issue to be discussed in this context concerns whether the division of competences between the General Assembly, as representative of all (member) states, and the Security Council, as mandated to undertake certain tasks requiring the consent at least of the five permanent members, would continue to make sense. If for example action would need to be taken on an urgent basis with respect to extra-terrestrial life, achieving consensus on how to act would be considerably easier amongst the 15 states members of the Security Council as compared to the 193 states members of the General Assembly. At the same time, its acceptability may be lessened to the same extent.

A related key issue then concerns whether the distinction between the five current permanent members of the Security Council as veto-holders and all other states makes the same sense in the context of twenty-first century actions *vis-à-vis* (the possibility of) extra-terrestrial life that it made in the context of the middle of the twentieth century. The United States, Russia, China, the United Kingdom and France have rather varied capabilities in terms of the search for extra-terrestrial intelligence, and from that perspective some other countries could lay equal or even better claims to such capabilities—not to mention the intergovernmental ESA, harnessing the technological and scientific expertise of the most important spacefaring nations in Europe.

At the end of the day, however, it should again be realized that the United Nations can only do what its member states generally speaking allow it to do.[38]

---

[37]See "Legal aspects of NEO threat response and related institutional issues", University of Nebraska-Lincoln, Final Report, 9 February 2010, § 4.

[38]The United Nations is, after all, still an *intergovernmental*, not a *supranational* organization of sovereign member states.

Even broadening interpretations of existing UN Charter concepts such as 'self-defense' and 'threats to peace and security', which would trigger certain legal consequences in a semi-automatic manner, require at least silent consent or lack of fundamental opposition from the more powerful states on earth, including the five veto-wielding powers of the Security Council.

In conclusion, once the international peace or a threat to it may be at stake, which at least covers one type of possible scenarios in the present context, the United Nations would probably be the 'least-worse' platform. More substantial questions would arise, however, once the discussion would extend to having the organization deal with *any* contact with extra-terrestrial life. It may well be that the United Nations is not yet ready for that, meaning that—for better or worse—individual states remain completely at liberty to handle such scenarios in a political and legal sense. One of those questions would concern whether the division of states as between a Security Council always comprising—next to ten temporary members—five permanent members with veto powers and the General Assembly comprising *all* states of the world does make as much sense here as it, apparently, does in the NEO context.

18.7 From Law to Meta-Law

It should be reiterated finally that the role of law is inherently limited to mankind. Humans namely also generally understand the unspoken underlying assumptions even if they do not always underwrite (all of) them. Extra-terrestrial life, obviously, presuming of course it possesses the requisite intelligence in the first place, does not necessarily share those assumptions, understandings or even the concept of 'law' as a binding set of specific social arrangements amongst humans—nor does it need to comply with it.

From this perspective, one should realize that there would be three generic scenarios at issue with respect to such extra-terrestrial life. Which of the three scenarios turns out to be the proper one would also fundamentally determine which role the United Nations, further to the above, would be likely to see thrust upon itself. The higher the perceived degree of hostility towards and vulnerability of mankind *vis-à-vis* any extra-terrestrial life would be, the more important it will be to address any such perceived threats in as unified a version as possible, where the United Nations again might then be seen as the 'least-worse' mechanism for achieving that.

Firstly, extra-terrestrial life might be fundamentally less intelligent and advanced than human life, in which case it likely has no concept such as 'law'. Should this mean we humans could treat them as animals, as objects of the law with rights but no self-standing capacity to stand up for them? Subject to certain decidedly non-legal or meta-legal phenomena such as ethics or enlightened self-interest which would likely intervene, this is at least probable to happen.

Secondly, extra-terrestrial life may be roughly as intelligent and advanced as human life, in which case it is likely to have a concept similar to law, a construct of broad binding arrangements on behaviour and its consequences designed to facilitate interaction and the realization of something like 'justice'—although it may be elaborated very differently. In such scenarios probably humanity should strive for a compromise 'meta-law', arranging essentially the respective spheres of application of human-made law and the comparable extra-terrestrial system, as well as finding some sort of common denominator, a compromise system or construct for inter-species social and other interaction.[39]

Under this scenario, an overarching 'regime' would thus be supposed to deal with the interaction between human life and extra-terrestrial intelligent life, by somehow providing each with their own domain, and adding a superstructure of meta-law. It would at least allow human life to continue applying the law amongst itself in relative independence. One might consider an approach furthermore of delineating the respective domains of application along 'physical'/'geographical' lines: humanity the solar system, the extra-terrestrials their own part of the universe, and in between 'cosmic oceans'.

Furthermore, such 'cosmic oceans' would preferably be roughly comparable to our terrestrial notion of 'global commons', essentially geographically delineated areas which cannot be appropriated, hence ruled legally, by any individual state or group of states and belong to society as a whole, whilst any individual state or group of states is entitled to use those areas at liberty, as limited only by general legal principles and laws.[40] This also raises a warning sign, as the concept of a 'commons' has also given rise to the 'tragedy of the commons', where all are entitled but no one (feels) responsible.

Thirdly, of course, extra-terrestrial life may be much more intelligent and advanced—in which case humanity is in trouble, at least as far as its laws are concerned. Whatever we might have concocted in terms of facilitating predictability and correctness of behaviour, it will depend on such extra-terrestrial life whether such a system and concept of 'law' will find any application *vis-à-vis* those extra-terrestrials, or even amongst ourselves. Humanity may end up being the 'object' of *their* system of 'law', or whatever has taken its place. This is essentially a legal version of the famous 'zoo hypothesis'—we are the intergalactic animals there, only objects of the superior observations, interests and resulting socio-politico constructs of superior creatures.

In sum, depending upon the level of intelligence and advancement of extra-terrestrial life, the foundations of the law will suffice, be thoroughly shaken in a need for a compromise, or found to be totally irrelevant in relation to such extra-terrestrial life—and perhaps elsewhere, too…

---

[39]In a sense, this mirrors the current role of international public law respectively private international law as, *inter alia*, accommodating the various national law regimes wherever international aspects or elements are involved, albeit at a conceptually higher level of course.

[40]Cf. for outer space Arts. II, I, Outer Space Treaty; and for the high seas Art. 87, United Nations Convention on the Law of the Sea, Montego Bay, done 10 December 1982, entered into force 16 November 1994; 1833 UNTS 3 & 1835 UNTS 261; UKTS 1999 No. 81; Cmnd. 8941; ATS 1994 No. 31; 21 ILM 1261 (1982); S. Treaty Doc. No. 103-39.

Erratum to: Shaking the Foundations of the Law: Some Legal Issues Posed by a Detection of Extra-Terrestrial Life

Frans G. von der Dunk

Erratum to: "Shaking the Foundations of the Law: Some Legal Issues Posed by a Detection of Extra-Terrestrial Life" in J.S.J. Schwartz and T. Milligan (eds.), *The Ethics of Space Exploration*, Space and Society, DOI 10.1007/978-3-319-39827-3_18

The copyright information given in the source line of Chapter 18 was not correct. The copyright holder of Chapter 18 is "© Frans G. von der Dunk 2016".

The updated original online version for this chapter can be found at 10.1007/978-3-319-39827-3_18

F.G. von der Dunk (✉)
College of Law, University of Nebraska-Lincoln, Lincoln, USA
e-mail: fvonderdunk2@unl.edu

J.S.J. Schwartz and T. Milligan (eds.), *The Ethics of Space Exploration*, Space and Society, DOI 10.1007/978-3-319-39827-3_19

Index

J.S.J. Schwartz and T. Milligan (eds.), *The Ethics of Space Exploration*, Space and Society, DOI 10.1007/978-3-319-39827-3

The manufacturer's authorised representative in the EU is Springer Nature Customer Service Centre GmbH, Europaplatz 3, 69115 Heidelberg, Germany. If you have any concerns regarding our products, please contact ProductSafety@springernature.com

Printed and bound by CPI Group (UK) Ltd, Croydon, CR0 4YY
30/09/2026
02225654-0011